BOB MILLER'S GEOMETRY FOR THE CLUELESS

GEOMETRY

Second Edition

BOB MILLER'S GEOMETRY FOR THE CLUELESS

GEOMETRY

Second Edition

Robert Miller

Mathematics Department
City College of New York

McGraw-Hill

New York   Chicago   San Francisco   Lisbon   London
Madrid   Mexico City   Milan   New Delhi   San Juan
Seoul   Singapore   Sydney   Toronto

3 4 5 6 7 8 9 0 DOC/DOC 0 9

ISBN 0-07-145902-2

Product or brand names used in this book may be trade names or trademarks. Where we believe that there may be proprietary claims to such trade names or trademarks, the name has been used with an initial capital or it has been capitalized in the style used by the name claimant. Regardless of the capitalization used, all such names have been used in an editorial manner without any intent to convey endorsement of or other affiliation with name claimant. Neither the author nor the publisher intends to express any judgment as to the validity or legal status of any such proprietary claims.

McGraw-Hill books are available at special quantity discounts to use as premiums and sales promotions, or for use in corporate training programs. For more information, please write to the Director of Special Sales, Professional Publishing, McGraw-Hill, Two Penn Plaza, New York, NY 10121-2298. Or contact your local bookstore.

Library of Congress Cataloging-in-Publication Data

Miller, Robert, 1943–
   Bob Miller's geometry for the clueless / Robert Miller.—2nd ed.
     p. cm.—(Bob Miller's clueless series)
   Includes index.
   ISBN 0-07-145902-2 (alk. paper)
1. Geometry. I. Title: Geometry for the clueless. II. Title.
  QA445.M512 2005
   516—dc22
                                             2005015297

To my dearest Marlene, I dedicate this book and everything else I ever do. I love you very, very much.

TO THE STUDENT

In many areas, geometry is much different from any math you may have had before or will see after this. So you must read the examples and proofs very slowly and carefully, most more than one time.

I have tried to make the examples and explanations as clear as I can. However, as much as I don't like to admit it, I'm not perfect. If you have questions, visit my Web site at www.mathclueless.com. You may also e-mail me at bobmiller@mathclueless.com.

Now enjoy the book and learn!

Bob Miller

CONTENTS

WHAT IS GEOMETRY? WHY SHOULD I TAKE IT?

The geometry we study, called *Euclidean geometry*, was written down about 2300 years ago in ancient Greece. It is an example of a *logic system*. At the start of every logic system are *undefined words*. Although Euclid had many more undefined words, we have narrowed the list down to six for convenience. Here they are:

1. Point

2. Line (infinite straight)

3. On (yes, indeed! *on*!!!)

4. Between

5. Congruent (identical in every way)

6. Plane (this one is usually undefined for convenience)

Notice what happens if we try to define the word *on* as "not off." We are forced to define the word *off*, which can be defined only as "not on." On and on and on. You see the problem. So *on* is undefined.

Next are the *defined words*. There are many. Let us list three. They will be repeated.

1. Line segment: All *points on a line between* two *points*.

2. Ray: All *points on* a *line on* one side of a *point on* that *line.*

3. Angle: Two *rays* with a common end*point.*

Notice all the undefined words in these three definitions. The last definition also uses a previously defined word.

Next there are *axioms* or *postulates,* laws taken to be true without proof. When I went to school, we used axioms for algebraic laws and postulates for geometric laws. I've continued that practice in this book, but other books may not do that. So you must check your book.

One postulate is that all radii of a circle are equal or congruent, depending on the book (you can't measure them all).

The last is a *theorem,* a law that is proven. One example is this: The sum of the measures of the angles of a triangle is 180°. We will need to know this fact for this course and the SAT.

Now that you have an idea what geometry is, you might ask, "Why should I take geometry?"

This is a very good question. Although it is necessary for students of math, it is just as important if not more so for nonmath students. Let me give you two examples.

A geometric proof is like a history essay. First you are given some facts and a question to answer. Then in a series of logical steps, you are to prove the answer, which is exactly a geometric proof without pictures.

Somewhere between your high school junior year and your college sophomore year, you may read Plato's *Republic.* Plato, a Greek philosopher, tried to prove what a perfect society would be. One of his undefined words was *good.* We all have an idea what a *good* person is. Plato then added defined words, axioms (postulates), and theorems, and he proved what a perfect society would be. Again, this is geometry without pictures. There are many more examples.

Finally you may ask, "Do I have a good book?"

For a book to be good, it must be clear and logical but not necessarily easy. Learning how to think can be compared to body building. Both are hard, but both are very satisfying after you have accomplished your purpose. Also, in general, the more nonmemorized-proof days you spend, the better the course. But we'll get very specific in the book.

You may notice that I have two definitions for some words. Also some theorems have two proofs: an old proof (before 1970) and a new proof. The problem with the newer proofs is that too much precision is sacrificed for understanding and logical thinking. Let me give you an example:

The picture in the margin is of a page from a book for a 6-year-old child.

The instruction says color the ball red.

The child picks up a red crayon and colors the ball red.

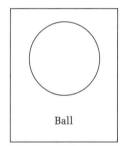

Ball

Buuuuuuut, the instruction is not precise.

The picture is not a ball but a picture of a ball.

The precise instruction is to color the *picture* of the ball red.

Because the instruction is too precise, the child does not know whether to color the ball or the whole page: more precision, less understanding.

I've tried to write this book so you'll understand either way!

O K A Y !!!!!!! Let's get going!!!!!!!!!

BOB MILLER'S GEOMETRY FOR THE CLUELESS

GEOMETRY

Second Edition

CHAPTER I

THE BASICS: UNDEFINED WORDS, DEFINED WORDS, AXIOMS, AND POSTULATES

UNDEFINED WORDS

At the start of geometry there are certain words that we cannot define, but we all know what they are. Let's look at them, explain them a little, and give their notation.

The first is *point.* Points are indicated with one capital letter. In the picture, A, B, and C are points:

The second is *line.* By a *line* we mean infinite straight line. We can use any of the following notations to describe the line in the picture: $\overleftrightarrow{AB}$, $\overleftrightarrow{AC}$, $\overleftrightarrow{BC}$, $\overleftrightarrow{BA}$, $\overleftrightarrow{CA}$, or $\overleftrightarrow{CB}$. All are correct.

The third is *on.* The point is *on* the line. Notice if you try to define *on* as "not off," then you have to define *off.* You go in circles.

The fourth is *between.* On the line, B is *between* A and C. Some books define *between.* Later we will give the definition.

The fifth is *congruent,* identical in every way. Its symbol is ≅.

The sixth is *plane*, flat and infinite in all directions. One way a plane is described is by listing three non-colinear points within the plane. An example will soon follow.

NOTE I
Colinear: Points all in the same line.
You must know these words verrry, verrry well.
Before we give an example, there are three kinds of lines we should mention:

    A. **Parallel lines:** Two lines that are everywhere equidistant. Two parallel lines are alllways in the same plane.

NOTE 2
Coplanar: In the same plane.

    B. **Intersecting lines:** Two lines that meet.

NOTE 3
Two intersecting lines are alllways in the same plane.

NOTE 4
Two lines can't meet at two points because it would only be one line.

    C. **Skew lines:** Two lines in space that are not parallel and don't intersect.

THE BEGINNING OF DEFINED WORDS

We are now starting to build up geometry. Let us define words that most of us know but that we need to know more precisely.

Line segment: All points on a line between two points.

A ————————————— B

The notation for our line segment is either AB or $\overline{AB}$. We will use $\overline{AB}$. Notice how many undefined words we used.

Ray: All points on one side of a point on a line.

A ————————————————→ B

The notation for a ray is $\overrightarrow{AB}$. This means all points starting from A, called the *endpoint,* through B. The ray $\overleftarrow{BA}$ is the ray with endpoint B through A.

←————————————— A ————— B

EXAMPLE 1

Here are the words for this picture.

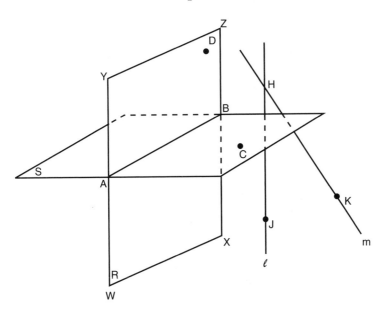

A. Give another description of plane S.

B. Give another description of plane R.

C. Give another description of line ℓ.

D. Give another description of line m.

E. If any other lines are pictured using two letters, name them.

F. Give some examples of lines that are not specifically drawn on the diagram.

G. Name parallel lines that are labeled on the picture.

H. Name intersecting lines that are labeled on the picture.

I. Name skew lines that are labeled on the picture.

J. Name rays on line ℓ.

ANSWERS—

A. Plane S is also plane ABC since A, B, and C are noncolinear points (not all in the same line).

B. There are many descriptions of plane R. Since A, B, D, W, X, Y, and Z are in this plane and since no three are colinear, any three would describe the plane—for example, XYZ or YXA or DWY or many others.

C. $\overleftrightarrow{HJ}$ or $\overleftrightarrow{JH}$.

D. $\overleftrightarrow{HK}$ or $\overleftrightarrow{KH}$.

E. $\overleftrightarrow{AB}$, $\overleftrightarrow{WX}$, and $\overleftrightarrow{YZ}$ are lines even though only part of them are pictured. There are others.

F. There are soooo many. Let me name a few: $\overleftrightarrow{BY}$ or $\overleftrightarrow{WZ}$ or $\overleftrightarrow{CX}$ or $\overleftrightarrow{DJ}$ or $\overleftrightarrow{DZ}$ or many, many, many more.

G. The same ones named in E, plus $\overleftrightarrow{WY}$ and $\overleftrightarrow{XZ}$.

H. l and m are intersecting lines. There are others pictured but none described using only one letter.

I. l or m, and any of the three lines in part E; it is hard to picture that these lines don't ever touch (they clearly are not parallel). We are trying to picture 3D on 2D paper; fortunately we don't have to do this much in the course.

J. $\overrightarrow{JH}$, the ray starting at J and going up, and $\overrightarrow{HJ}$, the ray starting at H, going down.

Let's look at angles.

Angle: Two rays or two line segments with a common endpoint.

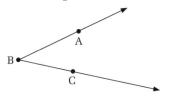

The common endpoint is called the *vertex* of the angle. Each angle pictured is read ∠ABC or ∠CBA. The vertex is allllways in the middle. The angle can also be written ∠B, by just naming the vertex only.

However, in the picture below, you cannot use only one letter since there are three angles with vertex B. Do you see them? They are ∠ABC, ∠CBD, annnd ∠ABD.

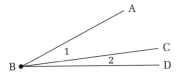

If I were very lazy, and I often am, I could have labeled ∠ABC as ∠1 and ∠CBD as ∠2 as in the picture. But do so only if your teacher allows it.

NOTE

The vertex is always the middle letter of the three letters.

∠ABC and ∠CBD are called *adjacent angles* because they are two angles with a common vertex and common side ($\overline{BC}$) between. ∠ABC and ∠ABD are *NOT* adjacent angles because the common side AB is not, not, not between the angles.

Many times in this book I will give you two definitions or two explanations. At the beginning, it will be because there are mainly two definitions used in many books. However, later, when we have other definitions or start proofs, I will give two proofs. One of them will be from when I took geometry. Since sometimes my kids call me "ancient," I will call them *ancient proofs*. I put those in because you will understand them better. The other will be called *newer*, and it might be in your book.

Warning!!!!!! Warning!!!!!! Every geometry book is different. You must use the statements that are in your book. This book is written for you to understand, but it may be slightly different from your book. *Warning!!!!!! Warning!!!!!!!!!*

ANGLES AND LINE SEGMENTS THE SAME

Ancient explanation: ∠A = ∠B. If you were to place angle A on top of angle B, they would appear to be the *same* angle. Although everyone understood this definition, according to the new math of the 1960s and 1970s (the worst thing to hit math until the new texts of the 1990s), this explanation is not technically correct.

Newer explanation: Physical objects, the angles themselves can only be *congruent.* ∠A ≅ ∠B. However, if both angles are 27°, their *measures* are equal. We write m∠A = m∠B. We then make the following definition.

DEFINITION: CONGRUENT ANGLES
$\angle A \cong \angle B$ if $m\angle A = m\angle B$. Similarly with line segments.

Ancient explanation: $\overline{AB} = \overline{CD}$. If you put $\overline{AB}$ on top of $\overline{CD}$, they would be exactly the same length, and you would see only one segment.

Newer explanation: $\overline{AB} \cong \overline{CD}$ if $m\overline{AB} = m\overline{CD}$. Fortunately most books don't do this. This is an inconsistency for which we are very happy.

A FEW POSTULATES AND SUPER IMPORTANT DEFINITIONS

Whenever anyone has trouble with geometric proofs, I always check this section first. You must know this section perfectly.

Postulate I Once around a circle is 360° (360 degrees). (Easy.)

Postulate 2a Line segments (ancient version): The whole is equal to the sum of its parts $\overline{AB} + \overline{BC} = \overline{AC}$. In some books (newer version): Definition of *between* is $\overline{AB} + \overline{BC} = \overline{AC}$ ($\overline{AB} + \overline{BC} \cong \overline{AC}$). Also, in some newer books, this is called *segment addition postulate*. Whew!!!

A   B   C

Postulate 2b Angles (ancient version): The whole is equal to the sum of its parts. $\angle ABC + \angle CBD = \angle ABD$. And (newer version): Angle addition postulate $m\angle ABC + m\angle CBD = m\angle ACD$.

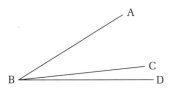

DEFINITION: MIDPOINT

M is the midpoint of $\overline{AB}$ if (ancient) $\overline{AM} = \overline{MB}$ or (newer) $\overline{AM} \cong \overline{MB}$. Alternate definition: (ancient) $\overline{AM}$ or $\overline{MB} = 1/2\ \overline{AB}$ or (newer) $\overline{AM}$ or $\overline{MB} \cong 1/2\ \overline{AB}$.

A ———————•——————— B
        M

DEFINITION: LINE SEGMENT BISECTOR

$\overline{CD}$ bisects $\overline{AB}$ at E if (ancient) $\overline{AE} = \overline{EB}$ or (newer) $\overline{AE} \cong \overline{EB}$.

Angle bisector: $\overline{BD}$ bisects angle ABC if $\angle ABD = \angle DBC$ ($\angle ABD \cong \angle DBC$).

In many geometry books, the definition of *perpendicular* is two lines or line segments that form equal (congruent) adjacent right angles. *You must check your book, for I have seen many different definitions for perpendicular, more than any other word.*

Perpendicular: Two lines (line segments) are perpendicular ($\perp$) if they form right angles, angles of 90°. On the picture if $\overline{AB} \perp \overline{CD}$, then rt $\angle ABC = (\cong)$rt $\angle ABD$ since perpendiculars form equal (congruent) right angles. In proofs, when we see perpendicular, we will write this statement out, usually leaving out rt (right).

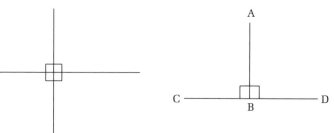

Complementary angles: Any two angles whose sum is 90° (sum of the measure of two angles is 90°). In symbols $\angle 1 + \angle 2 = 90°$ ($m\angle 1 + m\angle 2 = 90°$). The most common picture is below.

Note the spelling of *complementary* (complE . . .). When it is spelled *complimentary* (complI . . .), it is me telling you how good looking and smart you are (a compliment).

Supplementary angles: Any two angles whose sum is 180°. In symbols $\angle 1 + \angle 2 = 180°$ ($m\angle 1 + m\angle 2 = 180°$). The usual picture is below. Sometimes a pair of supplementary angles as pictured is called a *linear pair*.

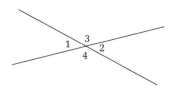

Vertical angles: As pictured, angles 1 and 2 and angles 3 and 4 are vertical angles.

NOTE

A linear pair is made up of two adjacent angles whose exterior sides are in a line. Just like other supplementary angles, $\angle 1 + \angle 2 = 180°$ ($m\angle 1 + m\angle 2 = 180°$).

Theorem (a statement to be proven): Vertical angles are equal (congruent). In symbols, $\angle 1 = \angle 2$ ($\angle 1 \cong \angle 2$). Also $\angle 3 = \angle 4$ ($\angle 3 \cong \angle 4$).

IMPORTANT IMPORTANT IMPORTANT IMPORTANT IMPORTANT IMPORTANT

If you do not know the definitions of *midpoint, angle bisector, segment bisector, complementary, supplementary*, and *vertical angles*, memorize and know them absolutely perfectly. If you don't memorize them, you will not be able to do the proofs.

I know more lists are boring, at least to some of you, but we need more facts.

SOME AXIOMS

Axiom 1: Addition   If equals (congruents) are added to equals (congruents), their sums are equal (congruent).

$x = 5$	$m\angle A = 32°$	$\angle C \cong \angle D$
$y = 2$	$m\angle B = 18°$	$\angle F \cong \angle G$
$x + y = 7$	$m\angle A + m\angle B = 50°$	$\angle C + \angle F \cong \angle D + \angle G$

Axiom 2: Subtraction   If equals (congruents) are subtracted from equals (congruents), their differences are equal (congruent).

$3x + 5y = 20$	$m\angle 1 = 57°$	$a \cong b$
$2x + 5y = 11$	$m\angle 2 = 23°$	$c \cong d$
$x = 9$	$m\angle 1 - m\angle 2 = 34°$	$a - c \cong b - d$

Axiom 3: Multiplication   If equals (congruents) are multiplied by equals (congruents), their products are equal (congruent).

$3x = 7$	$\dfrac{1}{4} AB = 5$	$\dfrac{1}{2} m\angle F = 27°$	$a \cong b$
$12x = 28$	$AB = 20$	$m\angle F = 54°$	$2a \cong 2b$

Axiom 4: Division   If equals (congruents) are divided by nonzero equals (congruents), their quotients are equal (congruent).

$12x = 36$	$\overline{AC} \cong \overline{FH}$
$x = 3$	$\dfrac{1}{3} \overline{AC} \cong \dfrac{1}{3} \overline{FH}$

Axiom 5: Reflective Law (In some books, it is called the *identity*.) A quantity is equal to (congruent to) itself.

$a = a$ $x + y = x + y$ $m\angle C = m\angle C$ $\overline{AB} \cong \overline{AB}$

Axiom 6: Symmetric Law (This is usually used to switch sides in an equation.) If $a = b$, then $b = a$. (If $a \cong b$, then $b \cong a$.) If $a + b = y$, then $y = a + b$. If $\overline{CD} \cong \overline{AB}$, then $\overline{AB} \cong \overline{CD}$.

Axiom 7: Transitivity (My favorite.) If $a = b$ and $b = c$, then $a = c$. (If $a \cong b$ aaaand $b \cong c$, theeeennnnn $a \cong c$.) If $x = y$ and $y = 7$, then $x = 7$. If $\overline{AB} \cong \overline{BC}$ and $\overline{BC} \cong \overline{CD}$, then $\overline{AB} \cong \overline{CD}$.

NOTE
There are other transitive laws. If $a < b$ and $b < c$, then $a < c$. The same is true for $\leq$, $>$, and $\geq$.

Axiom 8: Substitution (Technically it can be proven, but usually it is not. So we call it an *axiom*.)

   A. Quantities equal to (congruent to) the same quantity are equal to (congruent to) each other. If $a = x$ and $a = y$, then $x = y$. If $\overline{AB} \cong \overline{XY}$ and $\overline{AB} \cong \overline{YZ}$, then $\overline{XY} \cong \overline{YZ}$.

   B. Quantities equal (congruent) to equal (congruent) quantities are equal (congruent) to each other. If $a = x$, $b = y$, and $a = b$, then $x = y$. If $\angle A \cong \angle X$, $\angle B \cong Y$, and $\angle A \cong \angle B$, then $\angle X \cong \angle Y$.

   C. If a equals (is congruent to) b, then a may be substituted for b in any algebraic or geometric true statement. $a = b$. If $a + x = y$, then $b + x = y$. $\overline{AB} \cong \overline{CD}$. If $\overline{MQ} \cong 4\,\overline{AB}$, then $\overline{MQ} \cong 4\,\overline{CD}$.

There are more. Sorry. I really really am. But we must.

LAST DEFINITIONS AND POSTULATES BEFORE THE PROOFS

Acute angle: Any positive angle that is less than 90°.

Tight angle: That can't be correct. Ah! *Right angle:* A 90° angle.

Obtuse angle: An angle of more than 90° but less than 180°.

Straight angle: An angle of 180°.

Here are 10 postulates we are assuming. Later there will be some more.

Postulate 1 (again)   Once around a circle is 360°.

Postulate 2 (for the first time!)   One and only one line can be drawn between two points.

Postulate 3   Two lines cannot intersect at more than one point. (If they intersected at two points, by postulate 2 there would be only one line.)

Postulate 4   A line segment is the shortest distance between two points on a plane.

Postulate 5   Any geometric shape can be moved without changing its shape.

Postulate 6   A line segment has only one midpoint.

Postulate 7 An angle can have one and only one bisector.

Postulate 8 The perpendicular is the shortest line segment between a point and a line.

Postulate 9 The sum (of the measures) of the angles on one side of a line is 180°.

Postulate 10 Two adjacent angles are supplementary if their exterior sides are in a straight line. In some books this is called a *linear pair*.

I know most of these statements are obvious, but since we are not proving them or they cannot be proved, we have to list these postulates.

Now on to the beginnings of proofs???!!!!!!!!!!!!

CHAPTER 2

THE BEGINNINGS OF PROOFS

This is the reason you probably bought this book, so I'd better make this chapter great. Seriously, I hope you think all the chapters are great. I'm sure working hard enough for that to happen.

NOTE 1

Some of you probably started to read this book from here. You really should read *everything* that came before. It is verrry important.

NOTE 2

How good and valuable your geometry book is depends on how many weeks you spend working on original proofs. If it is half your year or more, you have a very good book and course. If you spend only a few weeks on proofs, you have a bad book. There is one crummy geometry book with only one chapter on proofs. Really bad!!!!!

Formal proofs are meant to train you in how to use logic, and they are very exact. Here are some beginning steps. More will follow as the proofs get longer. Notice

I didn't say harder. Once you get the hang of these, it won't matter how long.

1. In the upper left, draw the picture, a larrrge picture, with a ruler, even if the teacher doesn't. If the drawing is too sloppy or too small, you may not see what you should see!!!!

2. In the upper right, put the hypothesis (the facts you are given). In some theorems at the beginning, there might not be a hypothesis.

3. Under the given, write the conclusion, what you are trying to prove.

4. Draw a horizontal line under everything. Draw a vertical line about one third of the way on this line as long as you think the proof will take.

5. On the left are your statements you make.

6. On the right are your reasons the statements are true: undefined words, defined words, axioms, postulates, and previous theorems.

7. The last line under the statements should be exactly what you wanted to prove.

Last, we will do many proofs two ways: the ancient way (for more clarity) and a more recent way. Sometimes we will do the proofs together (to be brief). If we do, we will do the ancient way, and the newer way will be in (parentheses) as we did before.

Okay, okay!!!!!! Enough talking. Boy, I talk too much!!!! Let's do the first proof.

Theorem 1 Given line segment $\overline{AD}$ with points B and C between A and D. If $\overline{AB} = \overline{CD}$ ($\overline{AB} \cong \overline{CD}$), then $\overline{AC} = \overline{BD}$ ($\overline{AC} \cong \overline{BD}$).

Picture

A B   C D

What you should see: $\overline{AB}$ and $\overline{CD}$ are the same. Make sure when you draw the picture that they are the same!!! How do I get $\overline{AC}$? I add $\overline{BC}$ to $\overline{AC}$. How do I get $\overline{BD}$? I add $\overline{BC}$ to $\overline{CD}$. AHA!!! I am adding equals to equals.

It will take you a while to learn the style. But you can get it, and you *will* get it!!!!!!!!

EXAMPLE 1

A ⎯⎯⎯ B ⎯ C ⎯⎯ D

Given: 1. $\overline{ABCD}$ is a line segment
2. $\overline{AB} = \overline{CD}$ ($\overline{AB} \cong \overline{CD}$)

Prove: $\overline{AC} = \overline{BD}$ ($\overline{AC} \cong \overline{BD}$)

Statement	Reason
1. $\overline{ABCD}$ is a line segment	1. Given (1 and 2 are true because we said so).
2. $\overline{AB} = \overline{CD}$ ($\overline{AB} \cong \overline{CD}$)	2. Given (we now add $\overline{BC}$ to both).
3. $\overline{BC} = \overline{BC}$ ($\overline{BC} \cong \overline{BC}$)	3. Reflexive law or identity law (depending on the book).
4. $\overline{AB} + \overline{BC} = (\cong) \overline{BC} + \overline{CD}$	4. If =s ($\cong$s) are added to =s ($\cong$), their sums are =($\cong$).
5. $\overline{AB} + \overline{BC} =(\cong)\overline{AC}$ $\overline{BC} + \overline{CD} =(\cong)\overline{BD}$	5. The whole is equal to the sum of its parts or definition of between or segment addition postulate.
6. $\overline{AC} = \overline{BD}$ ($\overline{AC} \cong \overline{BD}$)	6. Substitution (step 5 into step 4).

The difficult part of geometry is the discipline of putting everything you do down!!!!!. For most people, this is verrry DIFFICULT in the beginning. Reread the proofs. Write them down. You will get it. It is a matter of time and sticking with it. Let's do a similar one.

Theorem 2 $\overline{ABCD}$ is a line segment. $\overline{AC} = \overline{BD}$ ($\overline{AC} \cong \overline{BD}$).

Prove $\overline{AB} = \overline{CD}$ ($\overline{AC} \cong \overline{BD}$). The picture is the same as the above. What you should see: Going from larger line segments to smaller line segments, you must subtract. However, the steps are in a slightly different order.

Given: $\overline{ABCD}$ a line segment
$\overline{AC} = \overline{BD}$ ($\overline{AC} \cong \overline{BD}$)

Prove: $\overline{AB} = \overline{CD}$ ($\overline{AB} \cong \overline{BD}$)

Statement	Reason
1. $\overline{ABCD}$ a line segment	1. Given.
2. $\overline{AC} = \overline{BD}$ ($\overline{AC} \cong \overline{BD}$)	2. Given.
3. $\overline{AC} = \overline{AB} + \overline{BC}$ ($\overline{AC} \cong \overline{AB} + \overline{BC}$) $\overline{BD} = \overline{BC} + \overline{CD}$ ($\overline{BD} \cong \overline{BC} + \overline{CD}$)	3. The whole equals the sum of its parts or definition of between or segment addition postulate.
4. $\overline{AB} + \overline{BC} = \overline{BC} + \overline{CD}$ ($\overline{AB} + \overline{BC} \cong \overline{BC} + \overline{CD}$)	4. Substitution (3 into 2).
5. $\overline{BC} = \overline{BC}$ ($\overline{BC} \cong \overline{BC}$)	5. Reflexive or identity law.
6. $\overline{AB} = \overline{CD}$ ($\overline{AB} \cong \overline{CD}$)	6. If =s ($\cong$) are subtracted from =s ($\cong$), their differences are = ($\cong$).

NOTE

Most teachers may allow you to take short cuts when you write the reasons, and I probably will too just to save space. But write out the whole reason. My high school teacher Miss Griswold made us write all the reasons out. I was not a natural at geometry, but she made me very good. She was the best math teacher I

ever had. Remember I have had a lot of math teachers!!!!!!!!!

Let's do a similar proof with angles.

Theorem 3

Given: ∠WOX = ∠YOZ (∠WOX ≅ ∠YOZ)

Prove: ∠WOY = ∠XOZ (∠WOY ≅ ∠XOZ)

What you should see: This theorem is very similar to
theorem 1, only with angles.

Statement	Reason
1. ∠WOX = (≅) ∠YOZ	1. Given.
2. ∠XOY = (≅) ∠XOY	2. Reflexive (identity).
3. ∠WOX + ∠XOY = (≅)∠XOY + ∠YOZ	3. If =s (≅s) are added to =s (≅), their sums are = (≅).
4. ∠WOX + ∠XOY = (≅)∠WOY	4. The whole =s the sum of its parts (∠addition postulate)
5. ∠XOY + ∠YOZ = (≅)∠XOZ	5. Same as 4.
6. ∠WOY = (≅)∠XOZ	6. Substitution (4 and 5 into 3).

Once you get the flow of the proofs, they will go much
easier.

You might try to do the next theorem on your own. It
is similar to theorem 2 but with angles like those in
theorem 3.

Theorem 4 Same picture. Given ∠WOY = (≅)∠XOZ.
Prove ∠WOX = (≅)∠YOZ.

The next proof shows what happens when the new math was put into effect. The proofs are different enough so we will do it twice.

Theorem 5a   Supplements (sups) of equal angles (or the same angle) are equal.

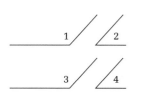

$\angle 1 = \angle 3$

Given: $\angle 1$ is the supplement of $\angle 2$

$\angle 3$ is the supplement of $\angle 4$

Prove: $\angle 2 = \angle 4$

Statement	Reason
1. $\angle 1$ is the sup of $\angle 2$	1. Given.
2. $\angle 1 + \angle 2 = 180°$	2. Definition of supplementary.
3. $\angle 3$ is the sup of $\angle 4$	3. Given.
4. $\angle 3 + \angle 4 = 180°$	4. Definition of supplementary.
5. $\angle 1 + \angle 2 = \angle 3 + \angle 4$	5. Substitution of 4 into 2.
6. $\angle 1 = \angle 3$	6. Given.
7. $\angle 2 = \angle 4$	7. If =s are subtracted from =s, their differences are =.

Theorem 5b

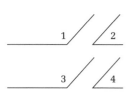

$\angle 1 \cong \angle 3$

Given: $\angle 1$ is the sup of $\angle 2$

$\angle 3$ is the sup of $\angle 4$

Prove: $\angle 2 \cong \angle 4$

Statement	Reason
1. $\angle 1$ is the sup of $\angle 2$	1. Given.
2. $m\angle 1 + m\angle 2 = 180°$	2. Definition of sups.
3. $\angle 3$ is the sup of $\angle 4$	3. Reason 1.
4. $m\angle 3 + m\angle 4 = 180°$	4. Reason 2.
5. $m\angle 1 + m\angle 2 = m\angle 3 + m\angle 4$	5. Substitution of 4 into 2.
6. $\angle 1 \cong \angle 3$	6. Reason 1.
7. $m\angle 1 = m\angle 3$	7. Definition of congruent angles.
8. $m\angle 2 = m\angle 4$	8. Subtraction (step 5 minus step 7).
9. $\angle 2 \cong \angle 4$	9. Reason 7.

There are a number of things to notice.

NOTE 1

Notice the way the proof flows. *Everything* has to be explained carefully (and written down) and logically. It is the way a lawyer builds a case to convict someone of a crime. The jury knows nothing. Everything is built up step by step until guilt is proven. The same holds for geometric proofs.

NOTE 2

Notice the given is not put down in one step and not necessarily in the order given. This is so the story is totally logical, from beginning to end.

NOTE 3

Notice the second way we proved this theorem we added to steps which made the proof a little longer and a little more difficult to understand and follow. Because of the new math distinguishing unnecessarily between physical angles and measures of angles, proofs have been made unnecessarily longer and pickier. But most modern books do it this way, so we are forced to do it.

NOTE 4
You should look at each theorem several times. Once you see the flow of the theorem, you will not have much trouble, if any.

Theorem 6   Complements of equal (congruent) angles or the same angle are equal (congruent).

The proof of this theorem is exactly the same as the proof of theorem 5 except . . . . the word *supplement* is replaced by the word *complement.*

aaaaaand 180° is replaced by 90°. The new picture is at the right. You should try the proof.

Let's do a couple of word problems. As you will see, geometry word problems are really easy. In fact the two here are probably the hardest possible, and they are easy, if your basic algebra is OK.

EXAMPLE 2
An angle is 4 times its complement. Find the angle.

Let x = angle. Thennnn 90 − x is its complement. The equation is

$x = 4(90 - x)$ $x = 360 - 4x$ $5x = 360$
$x = 72$ degrees $90 - x = 18$ degrees

Notice 4 times 18 is 72.

Let's try a slightly more complicated one, about as hard as you would get in all of geometry.

EXAMPLE 3—

Four times an angle is the supplement of its complement. Find the angle.

Let x = angle. 90 – x is the complement of the angle. 180 – (90 – x) is the supplement of the complement. Soooo

$4x = 180 - (90 - x)$ $4x = 90 + x$ $3x = 90$
$x = 30$ degrees

We still have some more introduction to do, believe it or not. We need to do a little formal logic, a very important part of geometry. Let's go on.

CHAPTER 3

IMPROVEMENT OF REASONING: STATEMENT, CONVERSE, INVERSE, CONTRAPOSITIVE; NECESSARY AND SUFFICIENT CONDITIONS; IF AND ONLY IF

Suppose we have the following statement:

If Sandy eats glecks, then Les sees flegs.

The part of the sentence after the word *if* is the **hypothesis**. It is the part that is **given**. The part of the sentence after the word *then* is the **conclusion**. It is the part to be **proved**.

The **converse** switches the *if* and the *then*.

Converse: If Les sees flegs, then Sandy eats glegs.

When the word *not* is in front of each part of the statement, it is called the **inverse**.

Inverse: If Sandy does **not** eat glecks, then Les does **not** see flegs.

The **contrapositive** both puts *not* in front of both parts of the statement and reverses.

Contrapositive: If Les does not see flegs, then Sandy does not eat glecks.

In logic, the statement and contrapositive are either both true or both false. Also the converse and inverse are either both true or both false. In good ol' English, this means you can prove either the statement or contrapositive and the statement is proven:

Proving the converse or inverse proves the converse true.

Let's give another example that makes this clearer.

Statement: A square is a rectangle.

Before the *is* is the hypothesis. After the word *is* is the conclusion.

Another way to write this statement is the following:

If a quadrilateral (four-sided polygon) is a square, it is a rectangle.

That is a true statement.

Converse: If a quadrilateral is a rectangle, then it is a square.

That is not true because the sides do not have to be the same.

Inverse: If a quadrilateral is not a square, it is not a rectangle.

That statement is false because it might be a rectangle.

Contrapositive: If a quadrilateral is not a rectangle, then it is not a square.

That statement is true, as we will look at in detail later.

NOTE

In this case both the statement and contrapositive are true, and the converse and inverse are both false.

Let's give one more example.

Statement: If ducks fly, then it will not rain.

Converse: If it will not rain, then ducks fly.

Inverse: If ducks do not fly, then it will rain.

Contrapositive: If it rains, then ducks do not fly.

NECESSARY AND SUFFICIENT CONDITIONS

Sufficient condition: We are given a fact or series of facts. If these facts make the conclusion true, then these are *sufficient conditions*.

EXAMPLE 1—

If I have a $10 bill, I can buy a 45-cent postage stamp.

EXAMPLE 2—

If a quadrilateral is a square, then the diagonals bisect each other.

In each case, the conditions are **sufficient** but are not **necessary**. I don't need a $10 bill to buy the stamp; less will do. A quadrilateral doesn't have to be a square; it need only be a parallelogram.

A *necessary condition* is a fact or set of facts that is absolutely necessary to make the conclusion true.

Notice that the converse of each statement is false (or not always true).

Consider the statement

If two lines are parallel, then alternate interior angles are equal (congruent).

Notice the statement is true, annnd the converse is true. That statement is a **necessary and sufficient condition**.

It can also be written as an **if and only if** statement, sometimes abbreviated by **iff**.

Two lines are parallel if and only if alternate interior angles are congruent (equal).

Statement: If two lines are parallel, alternate interior angles are equal (congruent).

Converse: If alternate interior angles are equal (congruent), then the two lines are parallel.

Please notice that necessary and sufficient are independent of each other.

EXAMPLE 3—

A condition that is both necessary and sufficient:

Statement: If the diagonals of a quadrilateral bisect each other, the figure is a parallelogram.

Sufficient: The statement itself is true.

Necessary (converse true): If a quadrilateral is a parallelogram, the diagonals bisect each other.

EXAMPLE 4—

A condition that is necessary but not sufficient:

Statement: If the diagonals of a quadrilateral are equal (congruent), the figure is a square.

Not sufficient: The figure could be a rectangle, an isosceles trapezoid, and a certain "arrowhead," as we will see later, as well as a square.

Necessary (converse true): If a quadrilateral is a square, then the diagonals **are** equal (congruent).

EXAMPLE 5—

A condition that is sufficient but not necessary:

Statement: If two angles are right angles, they are congruent.

It is sufficient because the statement is true.

Not necessary (converse false): If two angles are congruent, they are right angles.

EXAMPLE 6—

A condition that is both not necessary and not sufficient:

Statement: If a quadrilateral is a trapezoid, the diagonals are congruent (equal).

Both the statement and converse are false. The condition is not necessary and not sufficient.

NOTE

If you notice, this is more like your English class than your math class.

Remember, for most of you, the purpose of geometry is to improve your logical thinking and help you write better papers in English and history.

CHAPTER 4

PARALLEL LINES, FOREVER TOGETHER

This is a topic that most students like a whole lot. Let us define.

DEFINITION: PARALLEL LINES

Two lines *in a plane* that never meet. In space you can have nonparallel lines that never meet. They are called *skew* lines. Cross your arms without letting them touch each other. Your arms are skew. Let us list some facts about parallel lines.

Postulate 1   In a plane, lines are either parallel or they meet.

Postulate 2   Very, very important: Through a point in a plane not on a line, one and only one line can be drawn parallel to that line.

Fact 1   Parallelism *is not* reflexive: line a ∥ line a is false. A line cannot be parallel to itself.

Fact 2   Parallelism *is* symmetric: if a ∥ b, then b ∥ a.

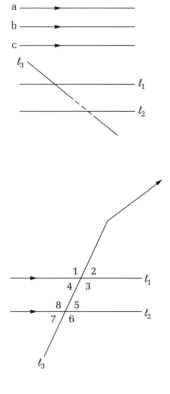

Fact 3   Parallelism *is* transitive: If a ∥ b and b ∥ c, then a ∥ c.

Fact 4   In a plane, if a line intersects one of two parallel lines, it intersects the other.

Facts 2 through 4 are small theorems and are not proved here.

We need a few definitions for parallel lines before we start.

l_3 is called a transversal; which is a line that cuts two or more other lines. We will use transversals to cut parallel lines.

l_1 and l_2 are parallel in this picture.

Angles 1 and 8 are called *corresponding angles.* In the two groups of angles they are in the upper left. So are 2 and 5 (upper right), 3 and 6 (lower right), and 4 and 7 (lower left).

Angles 3 and 8 are called *alternate interior angles* because they are on the opposite side of the transversal and between the lines. So are 4 and 5.

Angles 2 and 7 and 1 and 6 are called *alternate exterior angles.*

Angles 3 and 5 and 4 and 8 are *interior angles on the same side of the transversal.*

Angles 2 and 6 and 1 and 7 are *exterior angles on the same side of the transversal.*

The following theorems are very important.

Theorem 1   If two parallel lines are cut by a transversal, alternate interior (exterior) angles are equal (congruent).

Theorem 2   If two parallel lines are cut by a transversal, corresponding angles are equal (congruent).

Theorem 3 If two parallel lines are cut by a transversal, interior (exterior) angles on the same side of the transversal are supplementary.

As important, all the converses to these theorems are true, which, in general, does not happen.

Converse to Theorem 1 If two lines that are cut by a transversal have alternate interior (exterior) angles equal (congruent), then the lines are parallel.

Converse to Theorem 2 If two lines that are cut by a transversal have corresponding angles equal (congruent), the lines are parallel.

Converse to Theorem 3 If two lines that are cut by a transversal have interior (exterior) angles on the same side of the transversal supplementary, then the lines are parallel.

NOTE 1

A. To find alternate interior angles, look for the *letter Z or backwards Z.*

B. For corresponding angles, look for the *letter F or backwards F or upside down F.*

C. For vertical angles, look for the *letter X.*

D. For interior angles on the same side of the transversal, look for a *square C or backwards square C.*

NOTE 2

The SAT likes these questions. But you don't need to know them as well. You need to know that angles that look equal are equal. If they are not equal, they add to 180°.

NOTE 3

These theorems are proved by indirect proofs. We will do one of them later.

NOTE 4

In the picture, we have vertical angles that are the same: 1 and 3, 2 and 4, 6 and 8, 5 and 7.

NOTE 5

Teachers like to put numerical problems on parallel lines on tests. Let's do some now and save a few proofs for later.

The following picture below is for the next three examples.

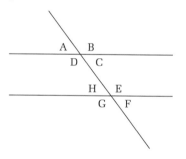

EXAMPLE 1

C is 4x − 2. ∠H is 2x + 24. Find x.

C and H are alternate interior angles. See the Z!!!

Sooooooo

$$4x - 2 = 2x + 24$$
$$+2 +2$$
$$4x = 2x + 26$$
$$-2x -2x$$
$$2x = 26$$
$$x = 13 \text{ degrees}$$

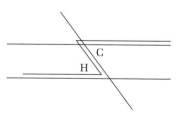

Angle C = 4(13) − 2 = 50 (degrees); and
H = 2(13) + 24 = 50 (degrees), as it must be.

EXAMPLE 2—

∠C is x + 5. ∠E is 4(x + 10). Find x.

C and E are interior angles on the same side of the
transversal. They are sups.

See the C.

$$x + 5 + 4(x + 10) = 180$$

$$x + 5 + 4x + 40 = 180$$

$$5x + 45 = 180$$

$$5x = 135$$

$$x = 27 \text{ degrees}$$

The angles are x + 5 = 32; 4(x + 10) =
4(27 + 10) = 4(37) = 148; 32 + 148 = 180, as it must.

EXAMPLE 3—

∠B is 7x − 2. ∠D is x + 88. Find x.

See the big X. B and D are vertical angles.

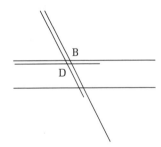

$$7x - 2 = x + 88$$

$$x = 15 \text{ degrees}$$

Both vertical angles are 103 degrees.

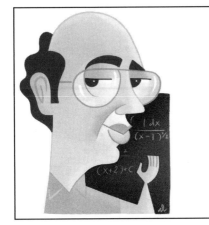

CHAPTER 5

MOSTLY TRIANGLES

This section is the beginning of the largest and most important part of the book. We will need some more definitions and examples.

DEFINITION: POLYGON

A closed connected figure made up of straight line segments. The figure below has five sides. It is called a *pentagon.* Each point is called a *vertex (plural is vertices).*

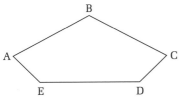

DEFINITION: QUADRILATERAL

A four-sided polygon. Examples are square, rectangle, trapezoid, rhombus, parallelogram, . . . We will look at these a lot later.

DEFINITION: TRIANGLE

A three-sided polygon. Triangles can be classified in two different ways: by sides or by angles.

Scalene triangle: A triangle with three unequal (noncongruent) sides and three unequal (noncongruent) angles.

The very popular *isosceles triangle:* A triangle with two equal (congruent) sides.

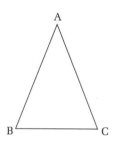

1. $\overline{AB}$, $\overline{AC}$ *legs:* the equal (congruent) sides.

2. $\overline{BC}$ *base:* usually not the same length as the legs, but might be. The base also doesn't have to be at the bottom, although it usually is.

3. ∠B and ∠C, the *base angles:* always opposite the legs; always equal (congruent).

4. ∠A, the *vertex angle:* always opposite the base; usually different from the other angles but not always; usually at the top of the triangle but not always.

Equilateral triangle: A triangle with all sides the same; it is also *equiangular;* each angle is 60° since, as we will soon show, the angles of a triangle total 180°.

Acute triangle: A triangle with all three angles less than 90°.

The most important one, *right triangle:* A triangle with one right angle.

1. ∠C is the *right angle,* denoted as in the picture.

2. ∠A and ∠B are acute angles. They usually are not the same.

3. $\overline{AB}$ is the *hypotenuse,* always opposite the right angle.

4. The hypotenuse is always the largest side since the largest side lies opposite the largest angle, and the largest angle lies opposite the largest side.

5. $\overline{AC}$ and $\overline{BC}$ are the legs that lie opposite the acute angles. They usually are unequal.

Obtuse triangle: A stupid triangle. Just kidding; a triangle with one obtuse angle.

Let's give examples of combinations of triangles.

Scalene and acute: angles are 50°, 60°, and 70°. The sides would also all be different.

Scalene and right: angles 30°, 60°, and 90°. We will see this one later.

Scalene and obtuse: angles 20°, 50°, and 110°.

Isosceles and acute: angles 50°, 50°, and 80°.

Isosceles and right: 45°, 45°, and 90°. We will see this one later, also.

Isosceles and obtuse: 20°, 20°, and 140°.

Buuut there is only one *equilateral* and *equiangular* triangle: 60°, 60°, and 60°.

Sorry, but we need a few more definitions. Remember this book has only the most essential definitions. At least one definition is a review.

Altitude: $\overline{AD}$ A line segment drawn from a vertex perpendicular to the opposite side and extended if necessary (as in the second figure).

> **NOTE**
>
> A triangle can't have two right angles since the third angle would have to be zero degrees.

> **NOTE**
>
> A triangle can't have two obtuse angles since we can't have two angles more than 90° and the angles of a triangle total only 180°.

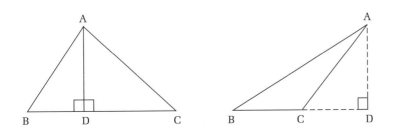

Angle Bisector (again) $\overline{BX}$ bisects $\angle ABC$ if $\angle ABX =$ ($\cong$)$\angle XBC$.

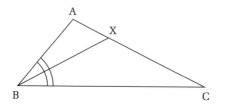

Median: $\overline{AY}$ A line segment that is drawn from a vertex to the middle of the opposite side. In other words, $\overline{AY}$ is a median if $\overline{BY} = (\cong)\overline{YC}$.

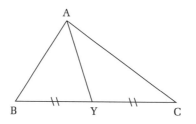

Exterior Angle $\angle AXY$ is an exterior angle. It is formed by extending one side of a triangle (or any other polygon). The angle on the outside of the triangle (or any other polygon) is that exterior angle. $\angle Y$ and $\angle Z$ are called *remote interior angles* to the exterior angle $\angle AXY$.

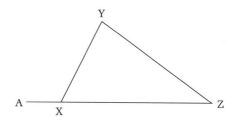

Next we will prove three theorems. The first two are vital not only for this course but also for the SAT.

Theorem 1 The sum of (the measures) of the angles of a triangle is 180°.

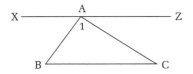

Prove: $(m)\angle B + (m)\angle 1 + (m)\angle C = 180°$

Statement	Reason
1. Draw line $\overleftrightarrow{XZ}$ through A $\parallel$ to $\overline{BC}$.	1. Through a point not on a line, one and only one line can be drawn $\parallel$ to a line.
2. $(m)\angle XAB + (m)\angle 1 + (m)\angle ZAC = 180°$	2. The sum of (the measures of) angles on one side of a line = 180°.
3. But $(m)\angle XAB = (m)\angle B$ aaand $(m)\angle ZAC = (m)\angle C$	3. If two lines are $\parallel$, (the measure of) alternate interior angles are =.
4. $(m)\angle B + (m)\angle 1 + (m)\angle C = 180°$	4. Substitution of 3 into 2.

Theorem 2 The (measure of the) exterior angle equals the sum of (the measure of) the remote interior angles.

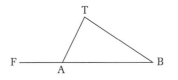

Prove: $(m)\angle FAT = (m)\angle T + (m)\angle B$

Statement	Reason
1. $(m)\angle FAT + (m)\angle TAB = 180°$	1. The sum of (the measures of) the angles on one side of a straight line is 180°.
2. $(m)\angle TAB + (m)\angle T + (m)\angle B = 180°$	2. Previous theorem. The sum of (the measures of) the angles of a triangle is 180°. (*Notice:* some theorems must be proven in sequence.)
3. $(m)\angle FAT + (m)\angle TAB = (m)\angle TAB + (m)\angle T + (m)\angle B$	3. Substitute 2 into 1.
4. $(m)\angle TAB = (m)\angle TAB$	4. Reflexive or identity.
5. $(m)\angle FAT = (m)\angle T + (m)\angle B$	5. Subtract 3 − 4.

Corollary   (a theorem, usually simple, which is the result of a more important theorem)   An exterior angle is larger than either of the remote interior angles since it is the sum of (the measures of) those two angles.

Here are the steps:

1. Draw the picture as before.

2. Write the given and prove as before.

3. Write all the possibilities.

4. In paragraph form, prove all but one of the possibilities wrong. The remaining one must be correct.

Theorem 3   If two lines form equal (congruent) alternate interior angles, the lines are parallel.

Given transversal t, ℓ_1, ℓ_2 lines: $(m)\angle x = (m)\angle y$

Prove: $\ell_1 \parallel \ell_2$

The only possibilities are that either the lines are parallel or the lines are not parallel. Suppose they are not parallel. Then the lines must meet in some point P and we form a triangle. But if this happens, $(m)\angle x > (m)\angle y$ since the exterior angle is larger than any remote interior angle. However, this is impossible because we assumed that $(m)\angle x = (m)\angle y$. Therefore assuming the lines are not parallel must be wrong. The only other possibility is $\ell_1 \parallel \ell_2$. A really good textbook has a chapter on proofs like this one.

Let's do a few numerical examples of theorems 1 and 2. They will probably be on your tests.

EXAMPLE I—

As you see, sides $\overline{AB}$ and $\overline{AC}$ are marked the same, so they are equal in length. Therefore the angles opposite are the same. Also the sum of (the measure of) the angles of a triangle $(\triangle) = 180°$. Soooo

$2x + 30 + 2x + 2x = 180 \qquad 6x + 30 = 180$
$6x = 150 \qquad x = 25 \qquad 2x = 50° \qquad$ and $\qquad 2x + 30 = 80°$

The three angles are $50°$, $50°$, and $80°$.

EXAMPLE 2

In a right triangle the acute angles total 90°. So

$$x - 12 + 2x + 3 = 90 \qquad 3x - 9 = 90 \qquad 3x = 99$$
$$x = 33 \qquad x - 12 = 21 \qquad \text{and} \qquad 2x + 3 = 69$$

Notice $21 + 69 = 90$.

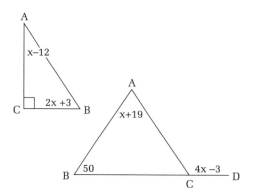

EXAMPLE 3

The (measure of) the exterior angle equals the sum (of the measures) of the remote interior angles.

$$50 + x + 19 = 4x - 3 \qquad x + 69 = 4x - 3 \qquad 72 = 3x$$
$$x = 24 \qquad x + 19 = 43 \qquad 4x - 3 = 93$$

Notice $50 + 43 = 93$.

We are now ready to begin the major proof section, the most important chapter or chapters in your book in terms of lasting benefits to you.

CHAPTER 6

SECRETS OF PROVING CONGRUENT TRIANGLES

Proofs like those in this chapter and in at least one later chapter determine how good a book is because these chapters truly teach you to think and write logically. If you spend half the course or more on doing mostly original proofs, then you have truly a value book to be treasured. If you spend only two or three or four weeks on memorized proofs, then either your book is worthless or it really isn't a geometry book (although it may look like a geometry book) or both. We do need one theorem before we get into the chapter.

Theorem Z If two angles of a triangle are = ($\cong$) to two angles of another triangle, the third angles are = ($\cong$).

Given: $(m)\angle A = (m)\angle X$

$(m)\angle B = (m)\angle Y$

Prove: $(m)\angle C = (m)\angle Z$

Statement	Reason
1. $(m)\angle A = (m)\angle X$, $(m)\angle B = (m)\angle Y$	1. Given.
2. $(m)\angle A + (m)B = (m)\angle X + (m)\angle Y$	2. Addition.
3. $(m)\angle A + (m)\angle B + (m)\angle C = 180°$	3. Sum of the (measures of the) $\angle$s of a $\triangle = 180°$.
4. $(m)\angle X + (m)\angle Y + (m)\angle Z = 180°$	4. Reason 3.
5. $(m)\angle A + (m)\angle B + (m)\angle C = (m)\angle X + (m)\angle Y + (m)\angle Z$	5. Substitution of 4 into 3.
6. $(m)\angle C = (m)\angle Z$	6. Subtract 5 minus 2.

Not too bad at all.

Now relax and read and do and reread and redo this next set of theorems. You will absolutely positively get the idea. But it may take some time!

There are four major methods of proving triangles congruent. These are theorems that I will not prove. Most books call them "postulates" so they will not have to prove them or explain why they don't prove them.

Theorem A: Side-Side-Side (abbreviation SSS)   If two triangles have three sides, of one equal (congruent) respectively to three sides of another triangle, then the triangles are congruent by side-side-side.
  Below are two such triangles. Equal (congruent) side (or angles) are marked the same.

Theorem B: Side-Angle-Side (SAS) If two triangles have two sides and the included angle of one is equal (congruent) = (≅) respectively (resp.) to two sides and the included angle of another triangle, then the triangles are congruent by side-angle-side.

Notice these triangles are backward.

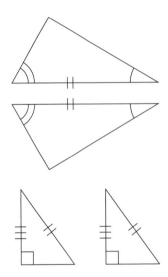

Theorem C: Angle-Side-Angle (ASA) If two triangles have two angles and the included side of one = (≅) resp. to two angles and the included side of another, then the triangles are ≅ by angle-side-angle.

Notice the triangles can be upside down.

Theorem D: Hypotenuse-Leg (HL) If a right triangle has its hypotenuse and one leg = (≅) resp. to the hypotenuse and leg of another right triangle, then the triangles are congruent by hypotenuse-leg.

It could have been the other leg also.

NOTE 1
There are other ways to show triangles congruent. These are all that are necessary. The others are just repeats, but we will mention them a little later.

NOTE 2
My geometry teacher was the best math teacher I ever had. She forced us to write out, using abbreviations, each reason for each step. I may not do it because it will make the book too long, but you should if you want to be very, very good at these proofs!!!!

NOTE 3

I think one of the worst things books do is spend weeks trying to get you to understand theorems A through D. As we do some of the proofs, you *will* get the idea very shortly.

We are going to do a number of proofs. We will show you how to write them down step by step. *More important,* we will show you how to *think through* the proofs so that you can do original ones easily on your own.

Theorem I Given isosceles triangle ABC, base $\overline{AC}$. $\overline{BD}$ bisects angle ABC. Prove triangle ABD is congruent to triangle CBD.

Usually the picture is given, especially in the beginning, but we redraw the picture with a ruler, and we put the given and what is to be proved as before. Here it is.

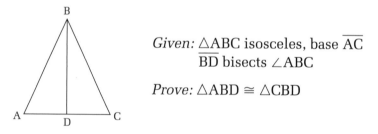

Given: △ABC isosceles, base $\overline{AC}$
$\overline{BD}$ bisects ∠ABC

Prove: △ABD ≅ △CBD

What You Should See First, you should see the triangles that are to be proven congruent. If you use a ruler, they will look the same. It really makes the proofs easier if the pictures are good.

   Second, look at the picture. You should be able to tell which parts of the triangles are the same iffff you had already proven the triangles congruent. $\overline{AB}$ is the same as $\overline{BC}$. Angle A and angle C are the same. Angle ABD and angle CBD are the same. $\overline{AD}$ and $\overline{DC}$ are the same. $\overline{BD}$ is in both triangles!!!

Third, in order to prove triangles congruent, we must use theorems A through D.

Last, we can only use the information given; not all the things eyes see are true.

Let's start.

The first step is to write out the first part of the given. The reason is given. Some books have the students write out the entire given in the first step. This is bad, bad, bad!!!!! It hurts understanding and the flow.

The first part looks like this:

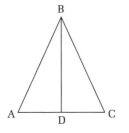

Given: △ABC isosceles, base $\overline{AC}$
$\overline{BD}$ bisects ∠ABC

Prove: △ABD ≅ △CBD

Statement	Reason
1. △ABC isosceles, base $\overline{AC}$	1. Given.
A. We have to tell what the uneven side is (the base).	
B. The next step is usually an equal (congruent) statement based on the previous step, and often the reason is explaining the previous step.	
The next line will be	
2. $\overline{AB} = \overline{CB}$ ($\overline{AB} ≅ \overline{CB}$)	2. Definition of isosceles triangle [a △ with two sides = (≅)].

Now the proof looks like this:

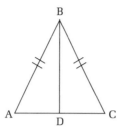

Given: △ABC isosceles, base $\overline{AC}$
$\overline{BD}$ bisects ∠ABC

Prove: △ABD ≅ △CBD

Statement	Reason
1. △ABC isosceles, base $\overline{AC}$	1. Given.
2. $\overline{AB} = \overline{CB}$ ($\overline{AB} \cong \overline{CB}$)	2. Definition of isosceles triangle.

We now start another sequence with the second part of the given. At the beginning the sequences will be only one or two steps, but soon there will be more. But they will come naturally.

Again, as before, mark the equal (congruent) parts the same. It looks like this:

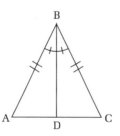

Given: △ABC isosceles, base $\overline{AC}$
$\overline{BD}$ bisects ∠ABC

Prove: △ABD ≅ △CBD

Statement	Reason
1. △ABC isosceles, base $\overline{AC}$	1. Given.
2. $\overline{AB} = \overline{CB}$ ($\overline{AB} \cong \overline{CB}$)	2. Definition of isosceles triangle.
3. $\overline{BD}$ bisects ∠ABC	3. Given.
4. ∠ABD = ∠CBD (∠ABD ≅ ∠CBD)	4. Definition of angle bisector.

What we have so far: We have side-angle. We need another angle before (for angle-side-angle) or a side after (for side-angle-side). But we have no more given!!!!!! What to do, what to do???!! Whenever there is no more information, we look for vertical angles (the letter X), a common side between the triangles, or, rarely, a common angle in overlapping triangles. Ahhhh, there is a common side. The triangles are congruent by side-angle-side. The finished proof looks like this:

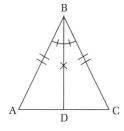

Given: △ABC isosceles, base $\overline{AC}$
$\overline{BD}$ bisects ∠ABC

Prove: △ABD ≅ △CBD

Statement	Reason
1. △ABC isosceles, base $\overline{AC}$	1. Given.
S 2. $\overline{AB} = \overline{CB}$ ($\overline{AB} \cong \overline{CB}$)	2. Definition of isosceles triangle.
3. $\overline{BD}$ bisects ∠ABC	3. Given.
A 4. ∠ABD = ∠CBD (∠ABD ≅ ∠CBD)	4. Definition of angle bisector.
S 5. $\overline{BD} = \overline{BD}$ ($\overline{BD} \cong \overline{BD}$)	5. Identity or reflexive.
6. △ABD ≅ △CBD	6. SAS. If two △s have two sides and the included ∠ of one = (≅) resp. to two sides and the included ∠ of the other, then the △s are ≅ by side-angle-side.

Comments

1. Read this proof over several times. Actually write it out. Try to get the sequence, the flow, the karma.

2. Mark S for side and A for angle to make sure you have everything, until you get good.

3. Read the proof.

4. Some books want the triangles, angles, and sides to be named in exactly the order the objects are the same. I don't care, but some books do. So check this out as you do the proofs.

5. Do number 3 again.

6. Do number 5 again until you understand it better.

Let's do another one, but we will not do it in such detail. Instead we will make a third column for comments.

Theorem 2

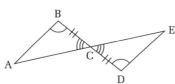

Given: C is the midpoint of $\overline{BD}$

$\overline{AB} \parallel \overline{ED}$

Prove: $\triangle ABC \cong \triangle BDC$

Statement	Reason	Comment
1. C midpoint of $\overline{BD}$	1. Given.	1. Write the first part of the given.
S 2. $\overline{BC} = \overline{DC}$ ($\overline{BC} \cong \overline{DC}$)	2. Definition of midpoint.	2. Explain the given.
3. $\overline{AB} \parallel \overline{ED}$	3. Given.	3. Write the second part of the given. When we have ∥ lines, we look first for the transversal. In this case, we see $\overline{BD}$. In this case we see the Z, for alternate interior angles (the Z is ABCDE). Angles B and D

Statement	Reason	Comment
		are the same. Notice there is another transversal $\overline{AE}$, a backward Z, BACED, and another set of alternate interior angles A and E, which we are not using in this proof but will in other proofs.
A 4. ∠B = ∠D ($\angle B \cong \angle D$)	4. If two ∥ lines are cut by a transversal, alternate interior angles are = (≅).	4. We now look for a common side or an X. This time we find the X for vertical angles!!!!
A 5. ∠BCA = ∠DCE ($\angle BCA \cong \angle DCE$)	5. Vertical angles are = (≅).	5. Nothing to add. Nothing? Impossible!! I talk too much!!!
6. △ABC ≅ △EDC	6. ASA. If two ∠s and the included side = (≅) resp. two ∠s and the included side of the other, the △s are ≅ by angle-side-angle.	6. The last statement should always look exacccctly like what you had to prove!!! Keep at it. You'll get it!!!!

Let's try another.

Theorem 3

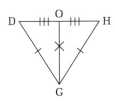

Given: $\overline{GO}$ bisects $\overline{DH}$

$\overline{GH} = \overline{GD}$ ($\overline{GH} \cong \overline{GD}$)

Prove: △DOG ≅ △HOG

Statement	Reason	Comment
1. $\overline{GO}$ bisects $\overline{DH}$	1. Given.	
S 2. $\overline{DO} = \overline{OH}$ ($\overline{DO} \cong \overline{OH}$)	2. Definition of segment bisector.	2. $\overline{GO}$ cuts $\overline{DH}$ in half.
S 3. $\overline{GH} = \overline{GD}$ ($\overline{GH} \cong \overline{GD}$)	3. Given.	3. A freebie.
S 4. $\overline{GO} = \overline{GO}$ ($\overline{GO} \cong \overline{GO}$)	4. Identity (reflexive).	4. Again a common side. Go go!!!
5. $\triangle DOG \cong \triangle HOG$	5. SSS. If three sides of one are = ($\cong$) resp. to three sides of the other, the $\triangle$s are $\cong$ by side-side-side.	5. This is an animal proof.

Getting better? Good. We'll do lots more.

Theorem 4

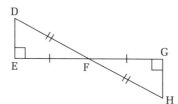

Given: $\overline{DE} \perp \overline{EG}$, $\overline{HG} \perp \overline{EG}$

F midpoints of both $\overline{EG}$ and $\overline{DH}$

Prove: $\triangle DEF \cong \triangle HGF$

Statement	Reason	Comment
1. $\overline{DE} \perp \overline{EG}$	1. Given.	1. Even if given is bunched, you separate.
2. DEF is a rt $\angle$	2. Definition of $\perp$.	2. $\perp$ is perpendicular (just making sure you know!).
3. $\overline{HG} \perp \overline{EG}$	3. Given.	3. We must delay = ($\cong$) because we only have one angle!
4. HFG is a rt $\angle$	4. Reason 2.	

Statement	Reason	Comment
rt ∠5. ∠DEF = (≅) ∠HGF	5. All rt ∠s are = (≅).	5. Whew! Lots of work for one part. Now that we have a right angle, we look for hypotenuse-leg.
6. F midpoint of $\overline{DH}$	6. Given.	
hyp. 7. $\overline{DF}$ = (≅) $\overline{FH}$	7. Definition of midpoint.	7. Here are our hypotenuses in both right triangles.
8. F midpoint of $\overline{EG}$	8. Given.	8. Another split of the hypothesis.
leg 9. $\overline{EF}$ = (≅) $\overline{GF}$	9. Reason 7.	9. Our legs.
10. △DEF ≅ △HGF	10. If two rt △s have the hyp. one leg = (≅) resp. to the hyp. and leg of the other, △s are ≅ by HL, hypotenuse-leg.	10. Hypotenuse-leg is only in a right triangle.

This proof is longer but not harder. Follow the flow, nice and easy.

LOTS OF NOTES

1. If you are very clever at this point, very, you might see that these triangles could have been proven congruent by SAS using vertical angles and totally avoiding the perpendicular. I did it this way for several reasons: (A) Show HL. (B) Show a slightly longer proof. (C) Point out that proofs with perpendicular tend to be a little longer.

2. While we are on perpendicular, let us see if it is reflexive, symmetric, or transitive. Let a, b, c stand for different lines:

   A. *Reflexive:* a ⊥ a. False. A line can't be perpendicular to itself.

B. *Symmetric:* If a ⊥ b, then b ⊥ a. True. See the picture below.

C. *Transitive:* If a ⊥ b and b ⊥ c, then a ⊥ c. False. In fact if you are in a plane, a would be *parallel* to c! See the picture above. If you were in space, it might be true. Look at the corner of your classroom. Orrrr. They might be skew lines, not parallel or perpendicular.
On a true-false test, "sometimes true is called FALSE"!

3. Triangles can be proven congruent by SAA or AAS (side-angle-angle or angle-angle-side). Since two angles are the same, the third angles must also (we proved). So we use ASA. However, if your book allows it, use SAA or AAS.

4. Some books have HA, hypotenuse-acute angle. Again, if you have two angles the same (acute angle and right angle), the third angles must be the same, and you can use ASA, which we will do. Again if your book allows it, use HA.

5. But triangles are *not, not, not* congruent by SSA, side-side-angle, the angle not between the two sides. Let's draw a picture to show two triangles not congruent if there is SSA.

Look at △ZXW and △YXW.

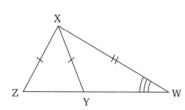

1. $\overline{ZX}$ is the same as $\overline{YX}$.

2. $\overline{XW}$ is a common side to both.

3. ∠W is common to both.

4. Side-side-angle of one = (≅) SSA, ∠W is not between the corresponding sides, in the others.

5. △ZXW is clearly not the same as △YXW.

Theorem 4 Revisited

Another form of proof that really can help you understand proofs is a **flow diagram**. It is now in a number of twenty-first-century textbooks. I believe it is an excellent idea to give you the flow of a proof, but I hope the teacher does not require these.

Let's do theorem 4 as a flow diagram.

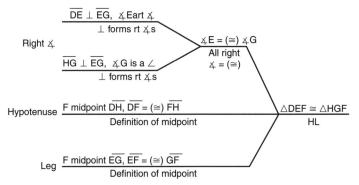

Another kind of proof is a **paragraph proof**. Since there are many in this book, we will not do theorem 4 as a paragraph proof.

Enough. Let's try another.

Theorem 5

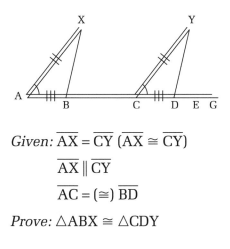

Given: $\overline{AX} = \overline{CY}$ $(\overline{AX} \cong \overline{CY})$

$\overline{AX} \parallel \overline{CY}$

$\overline{AC} = (\cong) \overline{BD}$

Prove: $\triangle ABX \cong \triangle CDY$

Thinking

1. We are given a freebie, $\overline{AX}$ is the same as $\overline{CY}$.

2. Parallel lines mean we need a transversal. It is ABCDEG. Here we have corresponding angles. See the F? It is crooked on its back.

3. $\overline{AC}$ and $\overline{BD}$ are too big and go outside the triangle. But subtracting $\overline{BC}$ from both, we get $\overline{AB}$ and $\overline{CD}$ the same.

4. We have side-angle-side. Ready? Let's write!

Statement	Reason
S1. $\overline{AX} = (\cong) \overline{CY}$	1. Given.
2. $\overline{AX} \parallel \overline{CY}$	2. Given.
A3. $\angle A = (\cong) \angle YCD$	3. If two $\parallel$ lines are cut by a transversal, corresponding $\angle$s are $=(\cong)$. Notice one angle needs only a single letter; the other angle three letters.
4. $\overline{AC} = (\cong) \overline{BD}$	4. Given.
5. $\overline{BC} = (\cong) \overline{BC}$	5. Reflexive (identity).
S 6. $\overline{AB} = (\cong) \overline{CB}$	6. Subtract 4 − 5.
7. $\triangle ABX \cong \triangle CDY$	7. SAS.

Theorem 6

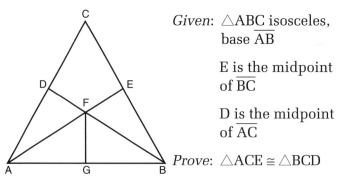

Given: $\triangle ABC$ isosceles, base $\overline{AB}$

E is the midpoint of $\overline{BC}$

D is the midpoint of $\overline{AC}$

Prove: $\triangle ACE \cong \triangle BCD$

Statement	Reason
1. $\triangle ABC$ isosceles, base $\overline{AB}$	1. Given.
S 2. $\overline{AC} = (\cong) \overline{BC}$	2. Definition of isosceles $\triangle$.
3. E is the midpoint of $\overline{BC}$ D is the midpoint of $\overline{AC}$	3. Given.
4. $\overline{CE} = (\cong) \frac{1}{2} \overline{BC}$ $\overline{CD} = (\cong) \frac{1}{2} \overline{AC}$	4. Definition of midpoint.
S 5. $\overline{CE} = (\cong) \overline{CD}$	5. Division or halves of $=$s ($\cong$s) are $= (\cong)$.
A 6. $\angle C = (\cong) \angle C$	6. Identity or reflexive law.
7. $\triangle ACE \cong \triangle BCD$	7. SAS.

LOTS OF NOTES

1. There are a lot of triangles in this picture. Make sure you draw this picture large, with a ruler.

2. If you cannot easily see which triangles you are proving congruent, outline both, each with a different color.

3. If you still can't see the triangles, draw them separately, again with a ruler.

4. Sometimes, in a proof, you may double up steps if the reasons are identical.

5. A picture similar to this one is really the only picture with an identity angle. Normally, you look for an identity side or vertical angles if not enough info is given.

6. Similar pictures (for identity angle) are line segment $\overline{CFG}$, $\overline{FG}$ missing, $\overline{AB}$ missing, or combinations of these.

Sometimes, in a proof, you must draw in **auxiliary** lines, extra lines, in order to complete a proof. Let us give an example of such a proof.

Theorem 7   If two sides of a triangle are equal (congruent), the angles opposite are equal (congruent).

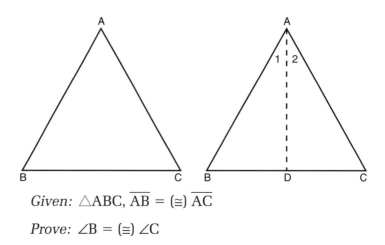

Given: △ABC, $\overline{AB}$ = (≅) $\overline{AC}$

Prove: ∠B = (≅) ∠C

Statement	Reason
S 1. $\overline{AB}$ = (≅) $\overline{AC}$	1. Given.
2. Draw $\overline{AD}$ bisecting ∠CAB	2. An angle can have one and only one angle bisector.
A 3. ∠1 = (≅) ∠2	3. Definition of angle bisector.
S 4. $\overline{AD}$ = (≅) $\overline{AD}$	4. Reflexive law or identity side.
5. △ADB ≅ △ADC	5. SAS.
6. ∠1 = (≅)∠2	6. CPCTE (CPCTC).

LOTS OF NOTES

1. This is an isosceles triangle theorem that says base angles are congruent.

2. The converse is true: If two angles of a triangle are congruent (equal), the sides opposite are congruent (equal).

3. If we had drawn $\overline{AD} \perp \overline{BC}$, we could have proved the triangles congruent by HL.

4. If we had drawn $\overline{AD}$ bisecting $\overline{BC}$, the triangles would be congruent by SSS.

5. If we had made D the midpoint of $\overline{BC}$ and then drawn $\overline{AD}$, again the triangles would have been congruent by SSS.

6. Sometimes in a proof, you might need more than one auxiliary line or line segment.

Let's do another proof that goes one step beyond congruent triangles. (Sounds like a good name for a sci-fi show.) The step after proving congruent triangles is called . . .

CORRESPONDING PARTS OF CONGRUENT TRIANGLES ARE EQUAL (CONGRUENT)

In order to prove congruent triangles, you need three select parts, each triangle to be the same. After you prove the triangles congruent, all remaining parts are the same. The reason will be corresponding parts. In virtually every proof (in your book, it might be every proof), after proving triangles congruent, the next reason will *always* be that corresponding parts of congruent triangles are equal (congruent), abbreviated by CPCTE (CPCTC). I do not like this because many students do not know and understand what this says by using the abbreviation, but I will use it later to save space.

Let's do it!! Yay!!

Theorem 8

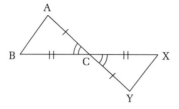

Given: $\overline{AY}$ and $\overline{BX}$ bisect each other

Prove: $\angle A = (\cong)\angle Y$

Statement	Reason
1. $\overline{AY}$ bisects $\overline{BX}$	1. Given.
S2. $\overline{BC} = (\cong)\ \overline{CX}$	2. Definition of segment bisector ($\overline{AY}$ is the cutter).
3. $\overline{BX}$ bisects $\overline{AY}$	3. Reason 1.
S4. $\overline{AC} = (\cong)\ \overline{CY}$	4. Reason 2 ($\overline{AY}$ has been cut).
A5. $\angle ACB = (\cong)\ \angle YCX$	5. Vertical angles are equal (congruent).
6. $\triangle ACB \cong \triangle YCX$	6. SAS. Even though in the proof we wrote SSA, in the picture the angle is between the sides. So we can use SAS.
7. $\angle A = (\cong)\ \angle Y$	7. Corresponding parts of congruent triangles are equal (congruent).

That's not so bad, is it?

When the proof goes beyond corresponding parts, we have to attack the proof differently. It's the way we *think* about the proof. You think about a proof *backward,* but you write it *forward.*

Let's try a few.

Theorem 9

Given: $\overline{VY} \perp \overline{WX}$

$\overline{VY}$ bisects $\angle$ WVX

Prove: $\triangle$ WVX is isosceles

Thinking   To show a triangle is isosceles, we must show two sides the same (definition). In this case we must show $\overline{WV}$ and $\overline{VX}$ the same. To do that we must have corresponding parts from congruent triangles. To prove congruent triangles, we need to show three parts the same. Let's do it!

Statement	Reason
1. $\overline{VY} \perp \overline{WX}$	1. Given.
A 2. $\angle$VWY = ($\cong$) $\angle$VYX	2. $\perp$ s form = ($\cong$) rt $\angle$s. Every book is slightly different. You must check your book, especially for this step, to make sure the reasons coincide.
3. $\overline{VY}$ bisects $\angle$WVX	3. Given.
A 4. $\angle$WVY = ($\cong$) $\angle$XVY	4. Definition of angle bisector.
S 5. $\overline{VY}$ = ($\cong$) $\overline{VY}$	5. Identity (reflexive).
6. $\triangle$WVY $\cong$ $\triangle$XVW	6. ASA. Two notes!!! Notice the side is between the angles; notice we didn't use HL because we didn't have the hypotenuse.
7. $\overline{WV}$ = ($\cong$) $\overline{VX}$	7. Corresponding parts of congruent triangles are equal (congruent). After congruent triangles always comes corresponding parts.
8. $\triangle$ WVX is isosceles	8. Definition of isosceles triangle (two sides the same).

That wasn't too bad. Let's try another. Remember, go over each example several times.

Theorem 10

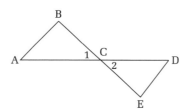

Given: C midpoint of $\overline{AD}$ and BE

Prove: $\overline{AB} \parallel \overline{DE}$

Thinking   In order to prove $\parallel$ we need to find a Z for alternate interior angles or an F for corresponding angles (others are rare). Here we see the Z (a little rhyme). To show the angles are the same we need corresponding parts after congruent triangles. For congruent triangles, we again need three proper things. In a very short time, this thinking will become natural. Also, in the real world, thinking should be done this way. Let's go!!!

Statement	Reason
1. C is the midpoint of $\overline{AD}$	1. Given.
S 2. $\overline{AC} = (\cong) \overline{DC}$	2. Definition of midpoint.
3. C is the midpoint of $\overline{BE}$	3. Given.
S 4. $\overline{BC} = (\cong) \overline{EC}$	4. Reason 2.
A 5. $\angle 1 = (\cong) \angle 2$	5. Vertical angles are equal (congruent).
6. $\triangle ABC \cong \triangle DEC$	6. SAS.
7. $\angle A = (\cong) \angle D$	7. CPCTE (CPCTC).
8. $\overline{AB} \parallel \overline{DE}$	8. If two lines cut by a transversal have alternate interior angles equal (congruent), the lines are parallel.

A student of mine asked a very intelligent question. She asked, "How many steps are in a proof?" It is a really good question.

1. Let's say we have to prove $\triangle ABC \cong \triangle XYZ$. The proof will take between four and nine steps. Usually it will take six or seven. Eight or nine steps is usually when you have perpendiculars. The final reason will be SSS, SAS, ASA, HL, or AAS (if your book allows this statement).

2. Let's say we have to prove $\angle 1 \cong (=)\angle 2$ or $\overline{AB} \cong (=) \overline{XY}$. There are two possibilities:

   A. It is one step more than in number 1 above. This is the much more likely. The last reason would be corresponding parts of congruent triangles are congruent (CPCTC) or corresponding parts of congruent triangles are equal (CPCTE).

   B. It could be a proof involving isosceles triangles. The length of these proofs varies greatly, from one step to more than nine steps.

3. Let's see what happens if you go beyond. Let's add a possible sequence of the last steps of a proof of each type.

   A. Two line segments parallel: $\triangle ABC \cong \triangle DCB$, $\angle ABC \cong (=)\angle BCD$ (CPCTC or CPCTE), $\overline{AB}\|$ $\overline{CD}$. [Usually the letter Z: If two lines cut by a transversal have alternate interior angles congruent (equal), then the lines are parallel, orrrr the letter F: If two lines cut by a transversal have corresponding angles congruent (equal), then the lines are parallel.]

   B. Two line segments perpendicular: $\triangle ABD \cong \triangle CBD$, $\angle ADB \cong (=)\angle CDB$ (CPCTC or CPCTE),

∠ADB and ∠CDB form a linear pair or are supplementary, AC ⊥ BD [if two line segments form equal (congruent) supplementary angles, the line segments are perpendicular.]
For all perpendicular and supplementary statements, you must carefully check your own textbook for the exact wording!!!!!!

C. A triangle is isosceles: △ABD ≅ △CBD, $\overline{AB}$ ≅ (=)$\overline{CB}$ (CPCTC or CPCTE), △ ABC is isosceles. (If two sides of a triangle are congruent, the triangle is isosceles.)

There are, of course, others. But this list covers the majority of the ones you get.

Now let's take a look at overlapping triangles.

Theorem 11: Overlapping Triangles

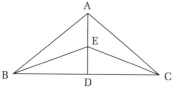

Given: △ABC isosceles, base $\overline{BC}$

   $\overline{AD}$ bisects ∠BAC

Prove: ∠AEB = (≅) ∠AEC

Thinking   We want to prove these angles the same. But these angles are in the upper two triangles. Our information is about △ABD and △ACD. So in order to show these angles the same by using corresponding parts, we must show the upper triangles congruent. To show the upper ones congruent, we must get corresponding parts from showing the big ones congruent. In each congruence, we must show three things, but some are the same in both. It's not as bad as it sounds, especially after you do seven or eight of them. (I said 7 or 8, not 70 or 80). *Hint:* If you have trouble seeing both sets of triangles, separate them!

Statement	Reason
1. △ABC isosceles, base $\overline{BC}$	1. Given.
S 2. $\overline{AB} = (\cong) \overline{AC}$	2. Definition of isosceles triangle.
3. $\overline{AD}$ bisects ∠BAC	3. Given.
A 4. ∠BAD = (≅) ∠CAD	4. Definition of angle bisector.
S 5. $\overline{AD} = (\cong) \overline{AD}$	5. Reflexive (identity).
6. △ADB ≅ △ADC	6. SAS.
7. ∠BAD = (≅) ∠CAD	7. CPCTE (CPCTC). Note ∠BAD and ∠BAE are the same angle!!! So are ∠CAD and ∠CAE.
8. $\overline{AE} = (\cong) \overline{AE}$	8. Identity (reflexive).
9. △AEB ≅ △AEC	9. SAS from steps 2, 7, and 8.
10. ∠AEB = (≅) ∠AEC	10. Corresponding parts of the upper triangles are equal (congruent).

As with all of these theorems, read them slooooowly, several times. Writing them out helps also. Boy am I tired from writing this chapter. You must be too. That's all for now. We'll revisit later.

CHAPTER 7

SIMILAR FIGURES AND PYTHAGORAS LIVES!!!!

Roughly speaking, similar figures look exactly the same, but are bigger or smaller. I don't think your teacher will let you get away with this explanation. So let's be more formal.

DEFINITION: SIMILAR

Two figures are similar if

    A. Corresponding angles are equal (congruent).

    B. Corresponding sides are in proportion.

Let us give an example.

NOTE

For this discussion, we will use the word equal only when we get to proofs (very few here).

EXAMPLE 1

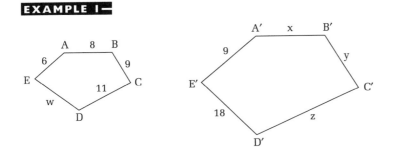

These two figures (pentagons) are similar iffffffffff

69

1. $\angle A = \angle A'$, $\angle B = \angle B'$, $\angle C = \angle C'$, $\angle D = \angle D'$, $\angle E = \angle E'$ annnnd

2. $\overline{AB} / \overline{A'B'} = \overline{BC} / \overline{B'C'} = \overline{CD} / \overline{C'D'} = \overline{DE} / \overline{D'E'} = \overline{EA} / \overline{E'A'}$

In solving for the missing letters w, x, y, and z, we find the corresponding sides in each figure that have numbers. In this case it is $\overline{AE}$ and $\overline{A'E'}$. $\overline{AE} / \overline{A'E'} = 6/9$ or 2/3 is called the *ratio of similarity*. When we compare any corresponding line segments in this figure, all will be in the ratio of 2/3. Once we say 2/3, it means you always start with the smaller figure.

We say $2/3 = w/18$. Cross-multiplying we get $3w = 36$. $w = 12$. Similarly $2/3 = 8/x$. $x = 12$. Also $2/3 = 9/y$. So $y = 13.5$. Lastly $2/3 = 11/z$. $z = 16.5$.

NOTE I
The notation for similarity is ~. Unfortunately, it is not on my computer; so I have to write them all in by hand.

NOTE 2
Congruent figures are similar, but similar figures don't have to be congruent.

NOTE 3
There are three ways to prove triangles similar:

A. *Angle-angle similarity:* If two angles of one △ are equal (congruent) to two angles of the other △, then the triangles are similar by angle-angle similarity.

B. *Side-angle-side similarity:* If two △s have consecutive sides in proportion and the included angles are equal (congruent), the triangles are similar by side-angle-side similarity.

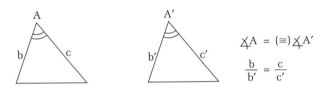

$$\angle A = (\cong) \angle A'$$

$$\frac{b}{b'} = \frac{c}{c'}$$

C. *Side-side-side similarity:* If two △s have their three sides in the same ratio, then the triangles are similar by side-side-side similarity:

$$\frac{a}{a'} = \frac{b}{b'} = \frac{c}{c'}$$

NOTE 4

We will do only a few (informal) proofs using similarity. Few schools do them anymore. In any case they are quite easy.

NOTE 5

In the case of similar figures in general, we need both angles the same and sides in proportion. In the case of triangles only, angles are enough (note 3A) or sides are enough (note 3C).

NOTE 6

We will do almost all similar proofs using angle-angle.

And finally,

NOTE 7

Recall that if two angles are the same, so are the third angles. Let's do another example.

EXAMPLE 2—

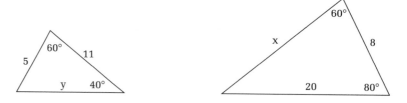

These triangles are similar since in the left triangle the missing angle is 80° and the missing angle in the second triangle is 40°, recalling the sum of the (measures of the) angles of a triangle is 180°. Also see that these triangles are backward. The ratio of similarity is 5/8.

To get the other ssssssidessssss:

$5/8 = y/20$ $8y = 100$ $y = 12.5$ aaand
$5/8 = 11/x$ $5x = 88$ $x = 17.6$

(Remember the side of the smaller triangle goes on top!)

EXAMPLE 3—

If $\overline{BC}$ is ∥ to $\overline{DE}$, we have similar △s, △ABC and △AED, which we will separate for clarity. You may want to do the same. The triangles are similar because of the corresponding angles ∠2 = (≅) ∠4 and ∠1 = (≅) ∠3. The △s are similar by angle-angle similarity.

8/14 or 4/7 is the ratio of similarity.

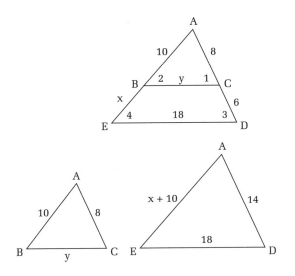

$4/7 = 10/(x + 10)$ $\qquad$ $4(x + 10) = 70$ $\qquad$ $4x = 30$

$x = 7.5$

(Little right side over big right side = little left side over big left side.)

$4/7 = y/18$

$\quad$ (Little right over big right = little base over big base.)

$7y = 72$ $\qquad$ $y = 72/7$

EXAMPLE 4—

Backward, upside down. $\overline{AB} \parallel \overline{ED}$ $\qquad$ $\triangle ABC \sim \triangle DEC$ since

$\angle A = (\cong) \angle D$ $\qquad$ and $\qquad$ $\angle B = (\cong) \angle E$

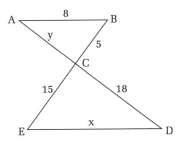

NOTE

In a proportion, you can also do it by saying little right over little base equals big right over big base. We will do this kind of proportion later. (You might try it here to show yourself it works.)

Alternate interior angles, and we again have angle-angle similarity.

$\overline{AC}$ corresponds to $\overline{DC}$!!! See??!! $\overline{BC}$ corresponds to $\overline{EC}$. Of course the bases also correspond.

$\overline{BC}/\overline{EC} = 5/15 = 1/3$, the ratio of similarity.

$1/3 = y/18$ $y = 6$ $1/3 = 8/x$ $x = 24$

This is an important figure that most books spend a few days on, and which leads to one of the most important results in geometry:

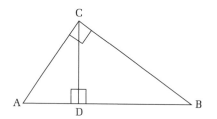

   Given right $\triangle ABC$ with the right angle at C. Drop the altitude to $\overline{AB}$ at point D. (Recall, altitudes form right angles.) There are three $\triangle$s here: small $\triangle DAC$, medium $\triangle DCB$, and large $\triangle CAB$ (sort of like papa bear, mama bear, and baby bear triangles).
   All three triangles are similar by angle-angle similarity. Why? First, all three triangles have right angles. Angle B is in the small triangle and the big triangle. So they are similar. Angle B is in the medium triangle and the big triangle. So all three triangles are similar!!!

In the small triangle DAC, the short leg is $\overline{AD}$, the long leg is $\overline{CD}$, the hypotenuse is $\overline{AC}$.

In the medium triangle DCB, the short leg is $\overline{CD}$, the long leg is $\overline{BD}$, the hypotenuse is $\overline{BC}$.

In the big triangle CAB, the short leg is $\overline{AC}$, the long leg is $\overline{BC}$, the hypotenuse is $\overline{AB}$.

In chart form, this is. . . .

	Short	Long	Hypotenuse
△DAC	$\overline{AD}$	$\overline{CD}$	$\overline{AC}$
△DCB	$\overline{CD}$	$\overline{BD}$	$\overline{BC}$
△CAB	$\overline{AC}$	$\overline{BC}$	$\overline{AB}$

Study this picture and the chart. Let's do a problem.

EXAMPLE 5—

Solve for x. Look for two △s where we can set up a proportion with numbers and the letter x only. We see the small and medium triangles fit the description. We say little short over little long equals medium short over medium long:

$2/x = x/8 \qquad x^2 = 16 \qquad x = 4$

Solve for y. Small and big triangles. Little short over big short equals little hypotenuse over big hypotenuse.

$2/y = y/(2 + 8) \qquad y^2 = 20 \qquad y = \sqrt{20} = 2\sqrt{5}$

Solve for z. We use the medium and large triangles. Medium long over medium hypotenuse equals big long over big hypotenuse.

$8/z = z/(2 + 8) \qquad z^2 = 80 \qquad z = \sqrt{80} = 4\sqrt{5}$

(If you need some help with the radicals, see the Appendix.)

This figure we are studying leads us into one proof of perhaps the most famous theorem in all of mathematics, the *Pythagorean theorem*.

If a and b are legs of a right triangle and the hypotenuse is c, then $c^2 = a^2 + b^2$.

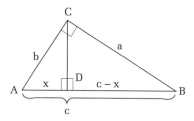

We have already shown these triangles to be similar. We let the letter c be the big hypotenuse $\overline{AB}$. If x is one part of the hypotenuse ($\overline{AD}$), the other piece, $\overline{BD}$, is c − x (the whole minus one piece is the other).

Using the little and big triangles and using small over hypotenuse in each, we get

$$x/b = b/c \qquad b^2 = cx$$

Using the middle and big triangles and using long over hypotenuse in each, we get

$$(c − x)/a = a/c \qquad a^2 = c(c − x)$$

Adding, we get

$$a^2 + b^2 = cx + c(c − x)$$

$$a^2 + b^2 = cx + c^2 − cx$$

The cx's cancel, and we get $a^2 + b^2 = c^2$ or $c^2 = a^2 + b^2$, using the symmetric law. Nice, isn't it? Believe it or not, there are over 100 more proofs of this theorem including three by past presidents of the United States. We actually had smart presidents who liked math.

Let's do a few examples.

EXAMPLE 6

$$a^2 + b^2 = c^2$$

$$5^2 + 7^2 = x^2$$

$$x^2 = 74$$

$$x = \sqrt{74}$$

EXAMPLE 7—

$a^2 + b^2 = c^2$

$x^2 + 8^2 = 9^2$ (hypotenuse is always alone)

$x^2 + 64 = 81$

$\qquad x^2 = 17$

$\qquad x = \sqrt{17}$

EXAMPLE 8—

$\qquad a^2 + b^2 = c^2$

$\qquad x^2 + (x + 7)^2 = (x + 8)^2$

$x^2 + x^2 + 14x + 49 = x^2 + 16x + 64$

$\quad 2x^2 + 14x + 49 = x^2 + 16x + 64$

$\quad -x^2 - 16x - 64 = -x^2 - 16x - 64$

$\qquad x^2 - 2x - 15 = 0$

$\qquad (x - 5)(x + 3) = 0$

So x = 5 and x = −3. −3 is a no no since sides can't be negative.

$x = 5$ $x + 7 = 12$ $x + 8 = 13$

The sides are 5, 12, and 13.

This brings us to the Pythagorean triples which you must memorize because they come up over and over and over and you will need them for this course, for the SAT, for trig and beyond, for science, for physics, . . . They are very necessary to speed your work.

Some of the triples are (the hypotenuse is last and always largest):

   3, 4, 5 group: 3, 4, 5 ($3^2 + 4^2 = 5^2$); 6, 8, 10; 9, 12, 15;
   12, 16, 20; 15, 20, 25.

5, 12, 13 group: 5, 12, 13; 10, 24, 26.

8, 15, 17 group: 8, 15, 17. That's all!

7, 24, 25 group: 7, 24, 25. That's all of them.

There are, of course, many others, actually an infinite number. Two more are 9, 40, 41 and 20, 21, 29. Should you learn these? No! They don't occur often enough. I've learned them because I want to. Math is such fun!!!!!!!!

There are two other right triangles of importance, one comes from a square and the other from an equilateral triangle.

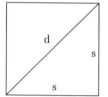

Draw the diagonal to the triangle. What you have is an isosceles right triangle. Since the right angle is 90°, the other two are 45° each, the 45°, 45°, 90° isosceles right triangle.

$$s^2 + s^2 = d^2 \qquad d^2 = 2s^2 \qquad d = s\sqrt{2}$$

In a 45°, 45°, 90° isosceles right triangle:

A. The legs are the same.

B. From the leg to hypotenuse, multiply by $\sqrt{2}$.

C. Hypotenuse to leg? Divide by $\sqrt{2}$. That's it.

EXAMPLE 9

$\overline{BC} = 6$

(legs the same)

$\overline{AB} = 6\sqrt{2}$ (leg times radical 2)

EXAMPLE 10

$$\overline{AC} = \frac{10}{\sqrt{2}} \times \frac{\sqrt{2}}{\sqrt{2}} = \frac{10\sqrt{2}}{2} = 5\sqrt{2}$$

$\overline{BC}$ and $\overline{AC}$ are the same. That's all there is to it.

Equilateral △ABC has its altitude (height) drawn. The two smaller triangles are congruent by HL (remember you can't forget the beginning). $\overline{BD}$ and $\overline{CD}$ are corresponding parts. Since all sides are "s," $\overline{BD} = 1/2$ s.

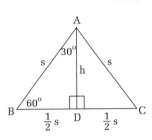

Using old Pythag

$$\left(\frac{1}{2}s\right)^2 + h^2 = s^2 \qquad \frac{1}{4}s^2 + h^2 = 1s^2 \qquad h^2 = \frac{3}{4}s^2$$

$$h = \frac{1}{2}s\sqrt{3}$$

Since all angles of △ ABC are 60° and, by corresponding parts, ∠BAD and ∠CAD are the same, both 30°.

Let us summarize:

Opposite the 90° angle, the hypotenuse is s.

Opposite the 30° angle, the short leg is $\frac{1}{2}$ s.

Opposite the 60° angle, the long leg is $\frac{1}{2}s\sqrt{3}$.

In rule form:

1. Always get the short leg first if it is not given!!!

2. Short to hypotenuse, multiply by 2.

3. Hypotenuse to short, divide by 2.

4. Short to long, multiply by $\sqrt{3}$.

5. Long to short, divide by $\sqrt{3}$.

EXAMPLE 11

Short (opposite the 30°) is given and is 7.

Hypotenuse is twice the short, 14.

Long is the short x$\sqrt{3} = 7\sqrt{3}$.

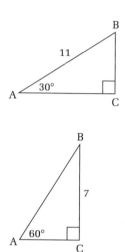

EXAMPLE 12—

Hypotenuse is 11.

Short is hypotenuse divided by 2: $11/2 = 5.5$.

Long is $5.5\sqrt{3}$.

EXAMPLE 13—

Long (opposite the 60° angle) is given: 7.

Short is long$/\sqrt{3} = 7/\sqrt{3} \times \sqrt{3}/\sqrt{3} = 7\sqrt{3}/3$.

Hypotenuse is twice the short $= 2 \times 7\sqrt{3}/3 = 14\sqrt{3}/3$.

NOTE I

When drawing a 30-60-90 triangle, always make the side opposite the 30 really look smaller than opposite the 60. It will really help. Always use a ruler!!!!

NOTE 2

We have shown the height of an equilateral triangle is $s\sqrt{3}/2$.

NOTE 3

The area of an equilateral triangle is $1/2(b \times h) = \dfrac{1}{2}s \times s\sqrt{3}/2 = s^2\sqrt{3}/4$. We will come back to this later.

Enough triangles already. Actually once you get by congruent triangles, the order of a book can be almost anything.

I'll tell you what. Let's do some quadrilaterals now.

CHAPTER 8

QUADRILATERALS SQUARELY DONE

Let's look at *quadrilaterals,* four-sided polygons. When we study them, you must learn their definitions and properties exactly. Teachers love to ask true-false and always-sometimes-never questions about these.

Parallelogram   A quadrilateral with opposite sides parallel. It has the following properties:

1. The opposite sides of a parallelogram are equal (congruent).

2. The opposite angles of a parallelogram are equal (congruent).

3. Consecutive angles of a parallelogram are supplementary.

4. One *diagonal* (a line segment drawn from one vertex to a nonconsecutive vertex) divides the parallelogram into two congruent triangles.

5. The diagonals of a parallelogram bisect each other.

How can you prove a quadrilateral is a parallelogram?

1. By showing opposite sides are parallel.

2. By showing both pairs of opposite sides are equal (congruent).

3. By showing one pair of opposite sides are parallel and equal (congruent).

4. By showing the diagonals bisect each other.

5. By showing the opposite angles are equal (congruent).

Notice: The second five are not all the converses of the first five.

You may have more proofs at this point. Let us do one.

Theorem I

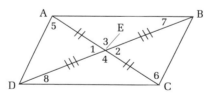

Given: $\overline{AC}$ bisects $\overline{BD}$

$\overline{BD}$ bisects $\overline{AC}$

Prove: ABCD is a parallelogram

Thinking   To show it's a parallelogram, we can show the opposite sides are parallel. We have two transversals $\overline{AC}$ and $\overline{BD}$ for alternate interior angles, the Zeeee. To get the angles the same, we'll get corresponding parts by proving two pairs of congruent triangles, the upper and lower, and the left and right. OK, let's do it.

Statement	Reason
1. $\overline{AC}$ bisects $\overline{BD}$ and $\overline{BD}$ bisects $\overline{AC}$.	1. Given (we're doubling up for brevity).
2. $\overline{DE} = (\cong)\ \overline{BE}$, $\overline{AE} = (\cong)\ \overline{CE}$	2. Definition of segment bisector.
3. $\angle 1 = (\cong)\ \angle 2$, $\angle 3 = (\cong)\angle 4$	3. Vertical angles are equal (congruent).
4. $\triangle AED \cong \triangle CEB$, $\triangle AEB \cong \triangle CED$	4. SAS in both cases.
5. $\angle 5 = (\cong)\ \angle 6$, $\angle 7 = (\cong)\ \angle 8$	5. CPCTE (CPCTC).
6. $\overline{AD} \parallel \overline{CB}$ and $\overline{AB} \parallel \overline{CD}$	6. If alternate interior angles are equal (congruent), the lines are parallel.
7. Quadrilateral ABCD is a parallelogram.	7. Definition of a parallelogram (opposite sides of a quadrilateral are parallel).

We will define the remaining quadrilaterals and list the properties, but your book may require you to prove some (not a bad idea since they are similar proofs and not too tough).

NOTE

It's a squashed square.

Rhombus   A parallelogram with all sides equal (congruent).

It has the same properties of a parallelogram + one.

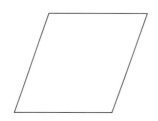

1. Opposite sides are equal (congruent) (actually all sides the same).

2. Opposite angles are equal (congruent).

3. Consecutive angles are supplementary.

4. One diagonal divides it into two congruent triangles.

5. The diagonals bisect each other.

The new kid on the block.

6. The diagonals are also perpendicular.

Rectangle   A parallelogram with right angles.

Properties 1 to 5 of a parallelogram and rhombus.

Not 6.

But 7. The diagonals are equal (congruent).

Square   A rectangle with all equal (congruent) sides. All properties 1 to 7.

Trapezoid   A quadrilateral with exactly one pair of parallel sides.

Isosceles Trapezoid   A trapezoid with nonparallel sides equal (congruent).

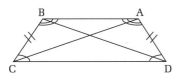

1. The base angles are equal (congruent), top and bottom.

2. The diagonals are equal (congruent).

$$\overline{AC} = (\cong) \ \overline{BD}$$

We introduce the following two figures because they will help you on true-false tests, and always-sometimes-never tests.

Kite   Quadrilateral with two pairs of nonopposite sides equal (congruent) and all angles less than 180°. The diagonals are perpendicular to each other.

Arrowhead   Quadrilateral with two pairs of nonopposite sides equal (congruent) and one angle more than 180°.

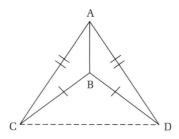

Properties:

1. The diagonals (extended if necessary) are perpendicular.

2. The diagonals *may* be equal (congruent).

NOTE
$\overline{AB}$ and $\overline{CD}$ are diagonals.

If you do not see the following picture about quadrilaterals in your book, the book is deficient.

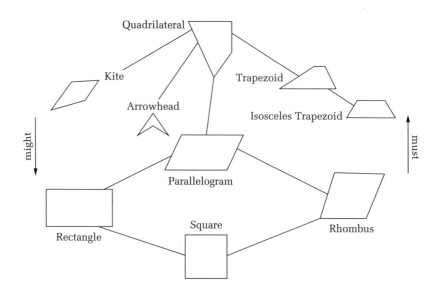

How to Read the Page   If a figure is a parallelogram, it must be a quadrilateral, but it might be a rectangle, rhombus, or square.

Must means true. *Might* means false on a true-false test or sometimes on an always-sometimes-never test.

Sample Question   A quadrilateral with two pairs of equal (congruent) sides is a parallelogram. Answer is sometimes (false on a true-false test). If the opposite sides of a quadrilateral are equal (congruent), it is *always* a parallelogram. This can be touugh. You really must understand and know each definition and each property of each figure. Too many *each's*. Let's go on.

CHAPTER 9

INTERIOR AND EXTERIOR ANGLES

Another topic, found in most math books and worth talking about, is interior and exterior angle of polygons. But first you should know, if you don't already, the names of polygons.

Once the numbers of sides gets to five or more, we will in general speak of *regular* polygons, polygons with all sides the same and all angles the same.

We of course know the regular three-sided figure: the *equilateral triangle*

And the regular four-sided figure: *square*

Pentagon: five-sided polygon

Hexagon: six-sided polygon

Heptagon: seven-sided polygon

Octagon: eight-sided polygon

Nonagon: guess . . . nine-sided polygon

Decagon: 10-sided polygon

Dodecagon: 12-sided polygon

Pentadecagon: 15-sided polygon

N-gon: n-sided polygon

NOTE
There are n(n − 3) /2 diagonals where n ≥ 3, n is the number of sides.

Now I do the problems on exterior and interior angles a little differently and a little easier than the standard book, so I'll do them both ways and you be the judge. First notice the following:

Three sides: one triangle

Four sides: two triangles with diagonals from one vertex

Five sides: three triangles

Six sides: four triangles

A. In general, the sum of the interior angles of any n-gon is (n − 2)180 degrees since there are (n − 2) triangles and the angles of each triangle add to 180 degrees.

B. An n-gon has n vertices. The sum of the interior plus the exterior angle is 180 degrees times n vertices is n(180) degrees. n(180) − (n − 2)180 = 2(180) = 360 degrees, the sum of the exterior angles.

C. One exterior angle is 360°/n where n is the number of sides.

D. The number of sides n is 360°/1 exterior angle.

E. One interior angle is (n − 2)180/n, but I'll try to show you easier ways.

Remember from the picture that the sum of an interior and an exterior angle is 180°.

In doing this kind of problem, *always* try to get the number of sides first. Always use the fact that the sum of the exterior angles is 360°.

EXAMPLE 1—

Tell me about a hexagon:

A. Six sides.

B. 360°/6 = 60°, each exterior angle.

C. 180° − 60° = 120°, each interior angle (the supplement of the exterior angle).

D. Total interior: 6(angles) × 120° (each interior angle) = 720 degrees   orrrr   4 triangles(6 sides − 2) × 180° (degrees in a triangle) = 720 degrees.

Not too bad. Let's try another.

EXAMPLE 2—

Each exterior angle is 10°. Find the rest.

A. Sides = 360/10 = 36 sides.

B. Interior angle is 180° − 10° = 170 degrees.

C. Total interior is (36)(170°) or 34(180°) = 6120 degrees. Messy but OK. Let's try one more.

EXAMPLE 3—

Total interior is 3240°. Tell me about the figure.

A. method 1: $(n - 2)\ 180 = 3240$

$$180n - 360 = 3240$$

$$180n = 3600$$

$$n = 20 \text{ sides} \qquad \text{orrrrr}$$

A. method 2: $3240/180 = 18$ triangles. 18 (triangles) + 2 = 20 sides.

B. $360/20 = 18$ degrees, one exterior angle. (The 18 is just a coincidence with the other 18.)

C. $180 - 18 = 342$ degrees, each interior angle. OK?? OK!!!

In addition, a polygon is called **concave** if you can find any two points inside the polygon, draw a line segment between those two points, and part of the line segment is outside the polygon.

If you cannot do that, then the figure is called **convex**.

Below are a concave figure and a convex figure.

Let's go on to areas and perimeters.

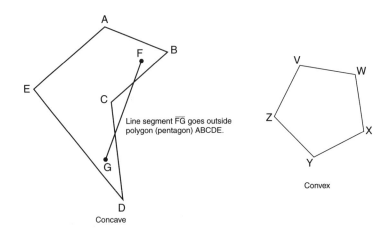

Line segment $\overline{FG}$ goes outside polygon (pentagon) ABCDE.

Concave

Convex

CHAPTER 10

AREAL SEARCH AND SECURING THE PERIMETER

This section is one you should probably be most familiar with, but let's make sure!

Rectangle

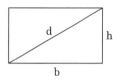

1. Area = base × height   A = bh.

2. Perimeter = 2b + 2h.

3. Diagonal d is found using old Pythagoras
   $d^2 = b^2 + h^2$.

EXAMPLE 1—

If the diagonal is 13 feet and the height is 5 feet, find the area and perimeter.

We must get the base first. Sooooo

$b^2 + h^2 = d^2$ $b^2 + 5^2 = 13^2$ $b^2 = 144$
$b = \sqrt{144} = 12$

But of course you knew this was a 5, 12, 13 right triangle.

$A = bh = 12 \times 5 = 60$ square feet (ft^2)

$p = 2b + 2h = 2(12) + 2(5) = 34$ feet

EXAMPLE 2

Find the area and perimeter of the following figure. All of the dimensions are in meters.

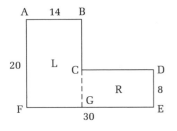

One way to get the answer is to extend $\overline{BC}$ until it hits $\overline{FE}$ at G, forming two rectangles.

$\overline{AB} = 14$ and $\overline{AB} = \overline{FG}$ $\overline{FG} + \overline{GE} = \overline{FE} = 30$

$\overline{GE} = \overline{CD} = 16$ $\overline{DE} = 8$ $\overline{DE} = \overline{CG}$

$\overline{CG} + \overline{CB} = \overline{AF} = 20$ So $\overline{BC} = 12$

Area of left rectangle = b × h = $\overline{FG} \times \overline{AF}$ = 14 × 20 = 280 square meters.

Area of right rectangle = b × h = $\overline{GE} \times \overline{DE}$ = 16 × 8 = 128 square meters.

Total area = 280 + 128 = 408 square meters (m^2)

Perimeter (add up all outside lengths) = $\overline{FE} + \overline{ED} + \overline{CD} + \overline{BC} + \overline{AB} + \overline{AF}$ = 30 + 8 + 16 + 12 + 14 + 20 = 100 meters.

Square

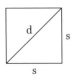

Rectangle where both base and height are the same.

$A = b \times h = s \times s = s^2$

$p = 2b + 2h = 2s + 2s = 4s$

$d = s\sqrt{2}$ as we showed in the 45-45-90 discussion.

EXAMPLE 3

Area of a square is 64 square inches. Find its perimeter and diagonal.

$A = s^2 = 64$ $s = \sqrt{64} = 8$ inches $d = s\sqrt{2} = 8\sqrt{2}$ inches $p = 4s = 4 \times 8 = 32$ inches

NOTE

Squaring comes from a square!!!! Guess where cubing comes from?

Triangle

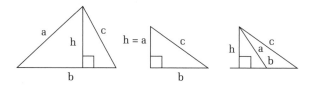

$$A = \frac{1}{2} \, bh \qquad p = a + b + c$$

h = height, is drawn perpendicular to one of the sides (base) and extended if necessary.

You might think a triangle is half a rectangle. You would be right!!!! Let's show it!

Both regions I and II are the same since the diagonal divides the rectangle (a parallelogram) in half. So a triangle really is half the rectangle. That's why the formula really is 1/2 bh for the area!!!

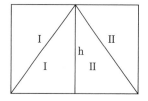

EXAMPLE 4—

In this right triangle, find a, the area, the perimeter, and y. All measurements are in centimeters.

1. $a^2 + 8^2 = 17^2$ $a^2 = 225$ $a = \sqrt{225} = 15$. Of course you knew this was an 8, 15, 17 right triangle.

2. A = 1/2 (b × h) = 1/2 (8 × 15) = 60 square centimeters (cm²).

3. p = 8 + 15 + 17 = 40 centimeters.

4. There are two ways to get the area: 1/2 (8 × 15) = 1/2 (17 × y), since y could be considered the height to base 17.

Sooo

120 = 17y   y = 120/17 centimeters

Trapezoid: Area = 1/2 h(b₁ + b₂)

Perimeter = ℓ_1 + b₁ + ℓ_2 + b₂

The perimeter is easy; add all the outside lengths. The area is the sum of the area of two triangles. As in the second trapezoid picture, the total area = 1/2 b₁h + 1/2 b₂h. Factoring out one-half h, we get A = 1/2 h(b₁ + b₂).

EXAMPLE 5—

Find the area and perimeter of the trapezoid if $\overline{AB}$ = 12, $\overline{AE}$ = 8, $\overline{BC}$ = 10 miles.

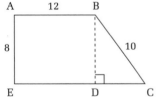

Draw $\overline{BD}$ perpendicular to $\overline{CE}$ making ABDE a rectangle. Triangle BDC is a right triangle.

$\overline{BC}$ = 10   and $\overline{BD}$ (= $\overline{AE}$) = 8

Soooo

$\overline{CD}$ = 6 (6-8-10 right triangle!)

$\overline{CE}$ = 12 + 6 = 18

ℓ_1 = h = 8   b₁ = 12 ℓ_2 = 10   and   b₂ = 18

p = 8 + 12 + 10 + 18 = 48 miles

A = $\dfrac{1}{2}$ 8(12 + 18) = 120 square miles (mi²)

Circle

Up to this point we have not dealt with the circle at all. We will deal with it now and several times in the rest of the book, in great detail. For now, let

O: the center of the circle.

c = *circumference:* perimeter of the circle.

r = *radius* (plural *radii*): distance from the center to any point on c.

d = *diameter:* distance from any point on c through O to c on the other side.

r = 1/2 d; d = 2r; c = 2πr or πd where π is approximately 3.14.

Pi is a number that originated with the Greeks. It is actually an irrational number, a nonrepeating and non-terminating decimal. Recently the most uninteresting book was published: 806 pages of pi written to 2 million places, 3.14159 . . . plus 1,999,995 more places. For the book even to be written, you get an idea of how important pi is in mathematics, verrry important!!!!!

Unfortunately, we cannot derive the formula for the area of the circle because it requires calculus. If you are interested, it can be found in *Calc II for the Clueless.*

EXAMPLE 6—

The diameter of a circle is 8 yards. Find its area and circumference.

$d = 8$ sooo $r = (1/2)8 = 4$

$A = \pi r^2 = \pi\, 4^2 = 16\pi$

I allow my students to leave it in this form, but many want you to do

$A = (3.14)(16) = 50.24 \text{ yd}^2$

$c = 2\pi r = 2(3.14)(4) = 25.12 \text{ yd}$

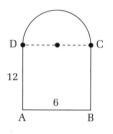

EXAMPLE 7—

We have a Norman window, with a semicircle sur-mounted on a rectangle. (This is the only time I ever see the word "surmounted.")

If $\overline{AB}$ = 6 feet and $\overline{AD}$ = 12 feet, find the area and perimeter.

First draw $\overline{CD}$. It is not part of the perimeter, but it helps you see the figure is a rectangle with a semicircle on top.

$\overline{AB} = \overline{CD}$ = 12 is the diameter of the semicircle. The semicircle r = 3.

$$A = \frac{1}{2}\pi r^2 + bh = \frac{1}{2}(3.14)(3^2) + (6)(12) = 86.13 \text{ square feet}$$

$$p = \overline{BC} + \overline{AB} + \overline{AD} + \text{a semicircle}$$

$$= 12 + 6 + 12 + (3.14)(3) = 39.42 \text{ feet}$$

Parallelogram

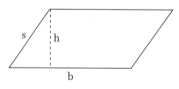

A = bh (the same as a rectangle). p = 2b + 2s.

Rhombus

$A = bh$ (since it is a parallelogram). Also $A = 1/2\, d_1 d_2$, half the product of the diagonal. $p = 4s$, since all sides are the same. Oh, let's do a problem.

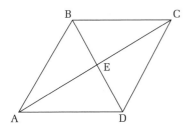

EXAMPLE 8—

Rhombus has sides 13 kilometers and the longer diagonal 24 kilometers, a big rhombus. Find the perimeter and area.

$p = 4s = 52$ kilometers

In the case of a rhombus, the diagonals not only bisect each other but are also perpendicular, forming right triangles. $\overline{AC} = 24$. $\overline{AE} = 12$, half of $\overline{AC}$. $\overline{AD} = 13$. Sooo $\overline{DE} = 5$, a 5-12-13 right triangle. $\overline{BD} =$ the other diagonal $= 10$.

$$A = \frac{1}{2}\, d_1 d_2 = \frac{1}{2}\,(24)(10) = 120 \text{ km}^2$$

Nice problem.

NOTE

If the *ratio of similarity* of two figures is ℓ_1/ℓ_2, the *ratio of their areas* is $(\ell_1/\ell_2)^2$.

EXAMPLE 9—

If the ratio of similarity of two triangles is 3/5, and the smaller area is 18 square units, find the area of the larger one.

The proportion is

$$\frac{18}{x} = \left(\frac{3}{5}\right)^2$$

$$\frac{18}{x} = \frac{9}{25}$$

The area of the larger triangle x = 50 square units.

Further note: If we had volumes, the *ratio of the volumes* would be $(\ell_1/\ell_2)^3$.

Sector (part of a circle)

AOB is a sector. Its area would be angle/360° πr^2 and the *arc length* s = angle/360° $2\pi r$, since it is that fractional part of the circumference.

EXAMPLE 10—

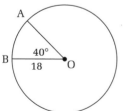

Let $\angle AOB = 40°$ and $\overline{OB} = 18$ inches. For the sector AOB, find its area, arc length, and perimeter.

$$A = \frac{40}{360}\, \pi\, (18)^2 = 36\pi \text{ square inches}$$

$$s = \frac{40}{360} 2\pi\, (18) = 4\pi \text{ inches}$$

$$p = s + 2r = 4\pi + 2(18) = 4\pi + 36 \text{ inches}$$

Equilateral Triangle

From before

$$A = \frac{s^2\sqrt{3}}{4} \qquad \text{and} \qquad p = 3s$$

EXAMPLE 11—

Suppose A = $16\sqrt{3}$, find p.

$$16\sqrt{3} = \frac{s^2\sqrt{3}}{4} \qquad \text{sooo} \qquad s^2\sqrt{3} = 64\sqrt{3} \qquad s^2 = 64$$

$s = 8 \qquad p = 3(8) = 24$

Finally let's talk about a *regular figure.*

The perimeter: the number of sides times s.

Area: Divide the figure into triangles.

The height of one of the triangles is called an *apothem.*

A = (1/2)(a)(s) times the number of sides. Butttt, s times the number of sides is the perimeter.

So

A = (1/2) *ap*

Enough of areas and perimeters. Let's do volumes and their surface areas.

CHAPTER 11

VOLUMES AND SURFACE AREAS

This is a topic that is largely ignored. A week or so should be spent on this chapter because the formulas are needed in later math courses and on the SAT (some of them).

Cube

The SAT is asking more about the shape. So we'll do the cube in great detail. (For more see *SAT Math for the Clueless*.)

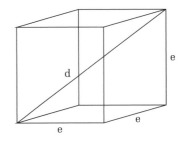

A cube has 12 edges (e), 6 faces, and 8 vertices.

The volume $V = e^3$, read, "e cubed." (Cubing comes from a cube!!!)

There are 6 faces (surfaces). Each is a square.

The surface area $S = 6e^2$.

The diagonal $d = e\sqrt{3}$.

EXAMPLE 1—

A cube has a volume of 64 cubic feet. Find its surface area and diagonal.

$$e^3 = 64 \qquad e = \sqrt[3]{64} = 4$$

$$S = 6e^2 = 6(4)^2 = 96 \text{ ft}^2 \qquad d = e\sqrt{3} = 4\sqrt{3} \text{ feet}$$

Box

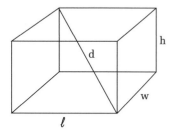

(The real name is *rectangular parallelepiped*, but that is just toooooo long!)

l: length   w: width   h: height

Volume $V = l \times w \times h$

Surface Area $S = 2lw + 2\,wh + 2lh$

(top and   (sides)   (front and
bottom)   back)

Diagonal $d = \sqrt{l^2 + w^2 + h^2}$, the 3-D Pythagorean theorem

Notice, the cube is a special case of a box.

EXAMPLE 2—

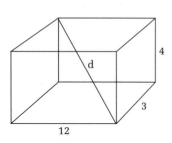

If the length of the box is 12, the width is 3, and the height is 4, find the volume, surface area, and diagonal (everything in meters).

$V = (12)(3)(4) = 144$ cubic meters

$S = 2(12)(3) + 2(3)(4) + 2(12)(4) = 192$ square meters

$d = \sqrt{12^2 + 3^2 + 4^2} = 13$ meters

Cylinder

Volume $V = \pi r^2 h$

Surface area $S = 2\pi r^2 + 2\pi rh$

The cylinder has three surfaces:

The top and bottom: $(2\pi r^2)$.

The curved surface: Take a can of soup. Take a knife and caaarefully cut the label straight up or down. You get a rectangle.

The height is the h of the can. The width is the circumference of the can 2π r. Area of rectangle = bh = 2πrh.

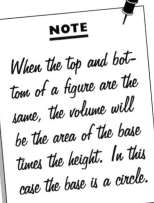

I once had a neighbor who wanted to know this formula. All he wanted to know was the answer, but as a teacher I had to tell him how to get it. That is why he moved. Just kidding.

EXAMPLE 3—

Find the volume and surface area of a cylinder whose diameter of the base is 20 feet and whose height is 7 feet. d = 20, r = 10.

$V = \pi (10)(10)7 = 700 \pi$ cubic feet

$S = 2\pi (10)(10) + 2\pi (10)(7) = 340\pi$ square feet

Sphere

$$V = \frac{4}{3} \pi r^3 \qquad S = 4\pi r^2$$

We need *Calc 1 for the Clueless* or *Calc 2 for the Clueless* for motivation and proof.

EXAMPLE 4—

Find the surface area if V = 36π cubic inches.

$$V = 36\pi = \frac{4}{3} \pi r^3$$

Canceling the pi on each side, and multiplying each side by 3/4, we get $r^3 = 27$. So

$r = 3$ and $S = 4\pi r^2 = 4\pi(3)^2 = 36\pi$ square inches

Notice the two numbers are the same. It is just a coincidence. One is a volume; the other is an area.

Cone

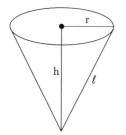

$$V = \frac{1}{3}\pi r^2 h \qquad S = \pi r^2 + \pi r \ell$$

ℓ is called the *slant height*.

NOTE
If the figure comes to a point, the volume is 1/3 area of the base times the height. Again the base is a circle.

EXAMPLE 5—

Find the volume and surface area of a cone if the radius of the base is 4 and the height is 6. All dimensions are in angstrom units.

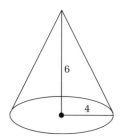

$$V = \frac{1}{3}\pi r^2 h = \frac{1}{3}\pi(4)^2(6) = 32\pi \text{ cubic angstrom units}$$

$$\ell = \sqrt{4^2 + 6^2} = \sqrt{52} = 2\sqrt{13}$$

$S = \pi r^2 + \pi r \ell = \pi(4^2) + \pi(4)(2\sqrt{13}) = 8\pi(2 + \sqrt{13})$ square angstrom units (the last step by adding and factoring)

Let's do two more figures.

EXAMPLE 6—

We have a prism (or a trough) whose ends are equilateral triangles. If the edge of the equilateral triangle is 10 feet and the length of the prism is 20 feet, find the volume and total surface area.

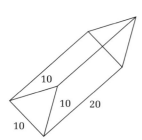

We can consider the equilateral triangle the base and the length the height of the figure.

Volume = area of the base times the height.

$$V = \frac{s^2\sqrt{3}}{4} \times h = \frac{10^2\sqrt{3}}{4}(20) = 500\sqrt{3} \text{ cubic feet}$$

There are five surfaces:

Two equilateral triangles: $S = 2 \times 10^2\sqrt{3}/4 = 50\sqrt{3}$.

Three rectangles = $3 \, b \times h = 3(10)(20) = 600$ square feet. Total area is $600 + 50\sqrt{3} = 50(12 + \sqrt{3})$ square feet.

EXAMPLE 7—

Given a right pyramid, square base, $\overline{AB} = 6$, height of pyramid $\overline{OP} = 4$. Find the volume V, slant height $\overline{PH}$, edge $\overline{PC}$, total surface area S. (A very popular problem.)

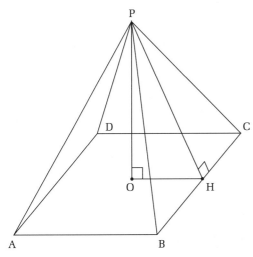

A right pyramid means the height of the pyramid is in the center of the base and of course perpendicular to the base ABCD.

Any figure coming to a point has its volume multiplied by 1/3. Since it is a square base the formula for the volume is

$$V = \frac{1}{3}s^2h = \frac{1}{3}6^2 \times 4 = 48 \text{ cubic units}$$

Since $\overline{AB} = 6$, $\overline{OH}$ which is half of $\overline{AB} = 3$.

$\overline{PH} = \sqrt{\overline{OH}^2 + \overline{OP}^2} = \sqrt{3^2 + 4^2} = 5$

A 3, 4, 5 Pythag triple again. If you look closely $\triangle CHP$ is a right triangle, where the angle at H is a right angle. $\overline{CH} =$ half of $\overline{BC} =$ half of $6 = 3$.

$\overline{CP} = \sqrt{\overline{CH}^2 + \overline{HP}^2}$

$\overline{CP} = \sqrt{3^2 + 5^2} = \sqrt{34}$

Total surface area is one base plus four triangular sides.

$S = s^2 + 4\frac{1}{2}b\ell = \overline{AB}^2 + 2(\overline{BC})(\overline{PH}) = 6^2 + 2(6)(5)$

$\qquad = 96$ square units

Now let's go back and look at the circle in greater detail.

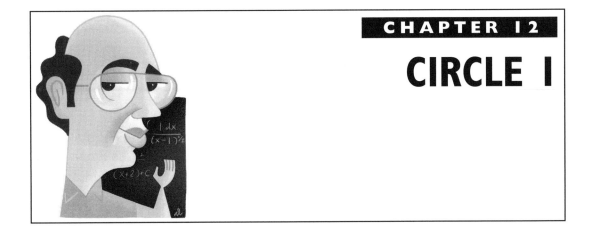

First let's take a look at the most common parts of a circle, reviewing from before and naming a few new parts.

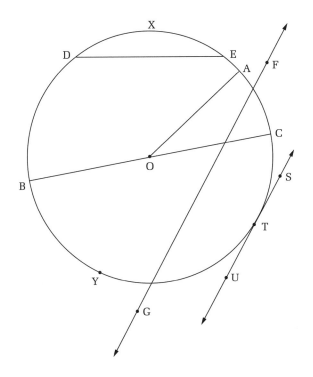

O: center of the circle.

$\overline{OA}$: *radius,* as before.

$\overline{BC}$: *diameter,* as before.

$\overline{DE}$: *chord,* a line segment from one side of the circumference to the other. The diameter is the largest chord.

$\overset{\frown}{DXE}$: *minor arc,* less than half a circle (*arc:* part of the circumference).

$\overset{\frown}{DYE}$: *major arc,* more than half a circle.

$\overset{\frown}{BDC}$: *semicircle,* half a circle.

$\overset{\leftrightarrow}{SU}$: *tangent,* a line hitting the circle at one and only one point.

T: *point of tangency.*

$\overset{\leftrightarrow}{FG}$: *secant,* a line hitting a circle in two places.

ARCS AND ANGLES WITH ARCS

Arcs can be measured in two ways. The first way, in terms of inches, meters, etc., we have already done. Arcs can also be measured in terms of degrees. As we have said, once around a circle is 360°. An arc that is 1/3 a circle is 1/3 of 360° = 120°.

Books today do almost no proofs in the area. So we will define a few more words, list the theorems, do some problems, and to make me feel a little better, I'll prove a theorem here and there.

DEFINITION I
Central angle, $\angle AOC$, an angle made with the vertex at the center, O, and whose sides are radii.

DEFINITION 2

Intercepted arc, $\overset{\frown}{AC}$*,* the arc that $\angle AOC$ cuts off on the circumference.

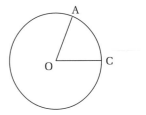

DEFINITION 3

The (measure of the) central angle is equal in degrees, $\overset{\circ}{=}$, to its intercepted arc. And just so I don't forget an important postulate . . .

Postulate I   All radii of the same or equal circles are equal (congruent).

DEFINITION 4

$\angle ABC$: an inscribed angle, an angle formed by two chords with its vertex on the circumference.

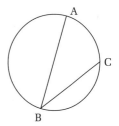

Theorem I   The (measure of the) inscribed angle $\angle ABC \overset{\circ}{=} 1/2 (m)\overset{\frown}{AC}$. Corollary: All inscribed angles in a semicircle are right angles!!!

Theorem 2   An angle formed by two chords that cross inside the circle is measured by half the sum (of the measures) of their intercepted arcs. Let's prove this one.

Given: $\angle 1$ formed by intersecting chords AB and CD at E, within the circle

Prove: $(m)\angle 1 \overset{\circ}{=} 1/2 \, [(m)\overset{\frown}{BD} + (m)\overset{\frown}{AC}]$

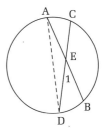

Thinking   Draw $\overline{AD}$. In $\triangle DEA$, $\angle 1$, an exterior angle is the sum of remote interior angles A and D. Each of the angles is half its intercepted arc. Now let's write it out.

Statement	Reason
1. Chords $\overline{AB}$, $\overline{CD}$ intersect at E	1. Given.
2. Draw $\overline{AD}$	2. One and only one line segment can be drawn between two points.
3. ∠1 exterior angle of △ADE	3. An exterior ∠ of a D is formed by extending one side of a triangle through a vertex.
4. $(m)\angle 1 = (m)\angle A + (m)\angle D$	4. The (measure of the) exterior angle is the sum of (the measure of) the remote interior angles.
5. $(m)\angle A \overset{\circ}{=} 1/2(m)\overset{\frown}{BD}$	5. The (measure of the) inscribed angle $\overset{\circ}{=}$ to (the measure of) one-half its intercepted arc.
6. $(m)\angle 1 \overset{\circ}{=} 1/2(m)\overset{\frown}{BD} + 1/2(m)\overset{\frown}{AC}$	6. Substitution of 5 into 4.
7. $(m)\angle 1 \overset{\circ}{=} 1/2((m)\overset{\frown}{BD} + (m)\overset{\frown}{AC})$	7. Factoring.

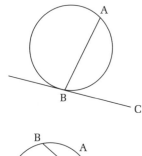

Theorem 3   The (measure of an) angle formed by a tangent and a chord to the point of tangency is equal in degrees to half its intercepted arc.

$$(m)\angle ABC \overset{\circ}{=} \frac{1}{2}(m)\overset{\frown}{AB}$$

Theorem 4   The angle formed by two secants intersecting outside the circle. The outside angle measures one-half the differences in the arc.

$$(m)\angle E \overset{\circ}{=} \frac{1}{2}[(m)\overset{\frown}{AB} - (m)\overset{\frown}{CD}]$$

Theorem 5   The angle formed by a secant and a tangent meeting outside the circle also measures one-half the difference of their intercepted arcs.

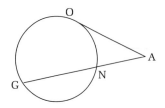

$$(m)\angle A \stackrel{\circ}{=} \frac{1}{2} [(m)\widehat{GO} - (m)\widehat{NO}]$$

Theorem 6   If two tangents meet outside the circle, that angle again measures half the difference of the intercepted arcs.

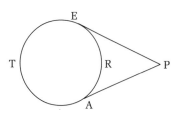

$$(m)\angle P \stackrel{\circ}{=} \frac{1}{2} [(m)\widehat{ATE} - (m)\widehat{ARE}]$$

In this case, since both arcs together make a circle, if we let x = the degrees of one of the arcs, the other would be 360 – x.

Let's try some problems. Most of the basic ones are verrry easy.

EXAMPLE 1—

$(m)\angle AOB = 42°$. Find $(m)\angle AXB$.

Angle AOB is a central angle and is equal in degrees to its arc AB. Angle AXB is half of its intercepted arc which is also AB. Half of 42 degrees is *21 degrees.*

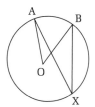

EXAMPLE 2—

Find (the measure of) x.

Angles that cross in the circle are half the sum of (the measures of) the arcs. $40° = (1/2)(x + 62°)$
$x + 62° = 80°$.

$x = 18°$

EXAMPLE 3

Find (the measure of) x.

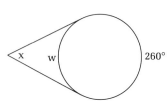

Any angle formed outside the circle measures half the difference in the arcs. But w = 100° = 360° − 260°.

$x = (1/2)(260° − 100°) = 80°$

EXAMPLE 4

Find (the measure of) x.

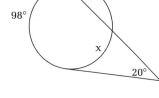

Again, the angle that is outside measures half the difference in the arcs.

$20° = (1/2)(98° − x)$ $40° = 98° − x$ $x = 58°$

Most of these basic ones are very easy. However in some books there are very long ones. If you do them systematically they are not too bad. Let's try one. Remember *don't panic!!!!!*

EXAMPLE 5

$(m)\overset{\frown}{LM} \cong 70°$ $(m)\overset{\frown}{JK} \cong 170°$ $(m)\angle UMJ = 40°$

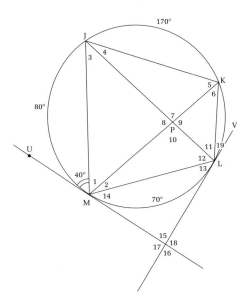

Find all the angles.

Don't panic!!!! It is not as bad as it looks. We first find all the arcs. (m)MJ $\overset{\circ}{=}$ 80° since it is double its inscribed angle UMJ (tangent and a secant). The remaining $\widehat{KL}$ = 360° − 70° − 80° − 170° = 40°. Now we can get all remaining angles

(m)∠1 $\overset{\circ}{=}$ 1/2(m)$\widehat{JK}$ = 1/2 (170°) = 85°, inscribed angle.

(m)∠2 $\overset{\circ}{=}$ 1/2(m)$\widehat{KL}$ = 1/2 (40°) = 20°, inscribed angle.

(m)∠3 $\overset{\circ}{=}$ 1/2(m)$\widehat{LM}$ = 1/2 (70°) = 35°, inscribed angle.

(m)∠4 $\overset{\circ}{=}$ 1/2(m)$\widehat{KL}$ = 1/2 (40°) = 20°, inscribed angle.

(m)∠5 $\overset{\circ}{=}$ 1/2(m)$\widehat{JM}$ = 1/2 (80°) = 40°, inscribed angle.

(m)∠6 $\overset{\circ}{=}$ 1/2(m)$\widehat{LM}$ = 1/2 (70°) = 35°, inscribed angle.

(m)∠7 $\overset{\circ}{=}$ 1/2(m)($\widehat{JK}$ + $\widehat{LM}$) = 1/2(170° + 70°) = 120°, intersecting chords in the circle   orrr
(m)∠7 + (m)∠4 + (m)∠5 = (m)∠7 + 20° + 40° = 180.
(m)∠7 = 120°, the sum (of the measures) of the angles of a triangle is 180 degrees.

(m)∠8 = (m)∠9 = 60°, both sups of angle 7.

(m)∠10 = 120°, vertical of angle 7.

(m)∠11 $\overset{\circ}{=}$ 1/2(m)$\widehat{JK}$ = 1/2 (170°) = 85°, inscribed angle.

(m)∠12 $\overset{\circ}{=}$ 1/2(m)$\widehat{JM}$ = 1/2 (80°) = 40°, inscribed angle.

(m)∠13 = (m)∠14 $\overset{\circ}{=}$ 1/2 (m)$\widehat{LM}$ = 1/2 (70°) = 35°, chord and tangent angle.

(m)∠15 $\overset{\circ}{=}$ 1/2(m)($\widehat{MJKL}$ − $\widehat{LM}$) = 1/2(290° − 70°) = 110°, two tangents meet outside the circle.

(m)∠16 = 110°, vertical of angle 15.

(m)∠17 = (m)∠18 = 70°, sup of angle 16.

(m)∠19 $\overset{\circ}{=}$ 1/2(m)$\widehat{KL}$ = 1/2 (40°) = 20°, chord and a tangent.

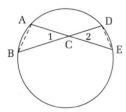

This really isn't so bad if you analyze slooooowly and carefully and don't panic. Problems like this occur in many geometry books, although most have fewer angles. Redraw this picture and try it yourself!!!!

There are certain lengths that can be proven. Like before, we will give the theorems, given examples, and a proof.

Theorem 7   If two chords cross, the product of each piece is the same.

Given: Chords $\overline{AE}$ and $\overline{BD}$ meet at C

Prove: $\overline{AC} \times \overline{CE} = \overline{BC} \times \overline{CD}$

Thinking   This is a cross-multiplication after finding corresponding parts. Corresponding parts are in proportion after proving similar triangles. The triangles are formed by drawing $\overline{AB}$ and $\overline{DE}$.

Statement	Reason
1. $\overline{AE}$, $\overline{BD}$ chords intersect at C	1. Given.
2. Draw $\overline{AB}$ and $\overline{DE}$	2. Through two points one and only one line segment can be drawn.
3. $(m)\angle B \stackrel{\circ}{\cong} 1/2(m)\overset{\frown}{AD}$	3. Inscribed angle = in degrees to half its intercepted arc.
4. $(m)\angle E \stackrel{\circ}{\cong} 1/2(m)\overset{\frown}{AD}$	4. Reason 3.
5. $\angle B = (\cong)\angle E$	5. Substitution (substitution + definition of congruent angles).
6. $\angle 1 = (\cong)\angle 2$	6. Vertical angles are equal (congruent).
7. $\triangle ABC \sim \triangle DEC$	7. Angle-angle similarity.
8. $\overline{BC}/\overline{CE} = \overline{AC}/\overline{CD}$	8. Corresponding parts of similar $\triangle$ s are in proportion.
9. $\overline{AC} \times \overline{CE} = \overline{BC} \times \overline{CD}$	9. Cross-multiplication orrr if equals are multiplied by equals, their products are equal.

EXAMPLE 6—

Solve for x. Kind of easy.

$6(x) = (3)(5)$ $6x = 15$ $x = 15/6 = 5/2$ or 2.5

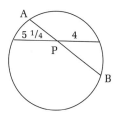

EXAMPLE 7—

If chord $\overline{AB}$ is 10, find each piece. Let x = one piece.
Then $10 - x$ is the other piece.

$$x(10 - x) = (5.25)(4)$$

$$10x - x^2 = 21$$

$$x^2 - 10x + 21 = 0$$

$$(x - 7)(x - 3) = 0$$

$$x = 3 \text{ and } 7 \text{ (the two pieces)}$$

Theorem 8 Given $\overline{AC}$ and $\overline{AY}$ secants meet at A,
then the products of the external piece and the whole
secant are equal.

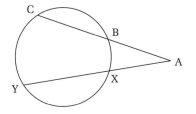

In symbols $\overline{AB} \times \overline{AC} = \overline{AX} \times \overline{AY}$. The proof is similar to
the last proof except for an identity angle instead of
verticals. Try to prove it!! I know you can!!

EXAMPLE 8—

Solve for x.

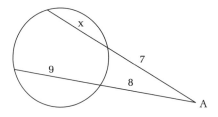

$$7(7 + x) = 8(8 + 9) \qquad 7x + 49 = 136 \qquad 7x = 87$$
$$x = 87/7 = 12.4 \text{ (approx.)}$$

Theorem 9   Given $\overline{AD}$ tangent and $\overline{AC}$ secant meet at A, then the tangent squared is equal to the product of the external piece and the whole secant. In symbols

$$\overline{AD}^2 = \overline{AB}(\overline{AB} + \overline{BC}).$$

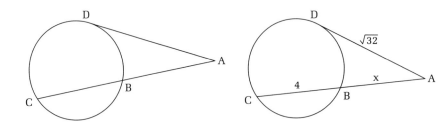

EXAMPLE 9—

Let $\overline{AC} = x \qquad \overline{BC} = 4 \qquad \overline{AD} = \sqrt{32}$. Find x.

$$x(x + 4) = (\sqrt{32})^2 \qquad x^2 + 4x - 32 = 0$$
$$x^2 + 4x - 32 = 0 \qquad (x + 8)(x - 4) = 0$$

$x = -8$ rejected since on this planet people can't be -8 feet tall.

The answer is $x = 4$.

OTHER THEOREMS

In some books they still do geometric proofs with circles. Hopefully your book still does them, and spends several weeks on them. If it does, the following is very important. In any case you should know the facts.

We will state a couple of postulates, state some theorems, and prove one.

Postulate 1 (again)   In a circle or equal circles all radii are equal (congruent).

Postulate 2 (new)   Any tangent is perpendicular to the radius drawn to the point of tangency.

Theorem 10   In a circle or equal circles, if any one of these is true, all are true.

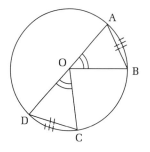

  A. Central angles are equal (congruent).

  B. Their chords are equal (congruent).

  C. Their arcs are equal (congruent).

Outline of Proof   Draw in the four radii $\overline{OA}$, $\overline{OB}$, $\overline{OC}$, and $\overline{OD}$. Assuming A or B or C is true, we can prove congruent triangles (differently in each case) to show the other two letters are true.

Theorem 11   In a circle or equal circles if any one is true, they all are true.

 A. The central angle is bisected.

 B. Its chord is bisected.

 C. The central angle's intercepted arc is bisected.

 D. The line segment is perpendicular to the chord.

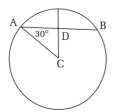

EXAMPLE 10—

Let $\overline{AB} = 12$ (m)$\angle A = 30°$ $\overline{CD} \perp \overline{AB}$.
Find the radius.

Since $\overline{CD} \perp \overline{AB}$, $\overline{CD}$ also bisects $\overline{AB}$.

Since $\overline{AB} = 12$, $\overline{AD} = 6$. $\triangle$ DAC is a 30-60-90 right triangle. 6 is leg opposite the 60° angle. To get $\overline{CD}$ the leg
opposite the 30° angle, we divide by square root of 3.

$$\overline{CD} = \frac{6}{\sqrt{3}} \times \frac{\sqrt{3}}{\sqrt{3}} = \frac{6\sqrt{3}}{3} = 2\sqrt{3}$$

Radius $\overline{AC}$ = hypotenuse = $2\ \overline{CD} = 4\sqrt{3}$.

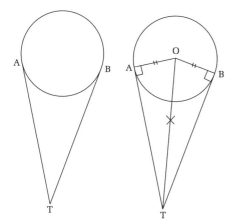

Theorem 12   Tangents from an external point to a circle are equal (congruent).

Given: $\overline{TA}$, $\overline{TB}$ tangents, circle center at O.

Prove: $\overline{TA} = (\cong)\overline{TB}$

Statement	Reason
1. $\overline{TA}$, $\overline{TB}$ tangents to circle O	1. Given.
2. Draw radii $\overline{OA}$, $\overline{OB}$, and line segment $\overline{OT}$	2. One and only one line segment can be drawn between two points.
3. $\overline{OB} \perp \overline{BT}$, $\overline{OA} \perp \overline{AT}$	3. Postulate: tangents drawn to the radii at point of tangency are perpendicular.
4. $\angle OBT = (\cong)\angle OAT$	4. Perpendiculars form equal (congruent) right angles.
5. $\overline{OA} = (\cong)\overline{OB}$	5. Postulate: All radii of a circle are equal (congruent).
6. $\overline{OT} = (\cong)\overline{OT}$	6. Reflexive (identity)
7. $\triangle OBT \cong \triangle OAT$	7. Hypotenuse-leg.
8. $\overline{TA} = (\cong)\overline{TB}$	8. Corresponding parts of congruent triangles are equal (congruent).

Notice the proofs are essentially the same, except you use the new properties of circles.

Unless geometry changes back to what it should, in which case there will be a new edition, these are all the congruent triangle proofs for now.

CHAPTER 13

LINES (THE STRAIGHT KIND) AND PARABOLAS I (NOT STRAIGHT)

This chapter is the start of a number of excellent sections (plus one bad one): lines, parabolas, ellipses, hyperbolas, and functions. The only thing is that they shouldn't be in geometry books. They should and must be in algebra books. However, since many geometry books include these topics, we will also.

In elementary algebra, you learned how to graph points and how to graph lines. We will briefly review graphing and then attack the harder problem, finding the equation of a straight line.

STANDARD FORM

When we look at an equation, we should instantly know that it is a straight line. Any equation in the form $Ax + By = C$, where A, B, and C are numbers and A and B are not both equal to 0, is a straight line.

NOTE

In some books, standard form is $Ax + By + C = 0$, so standard form isn't standard. Isn't that funny?!

Let's make sure we understand this by giving some examples that are straight lines and some that are not.

EXAMPLE I—

A. $3x - 4y = 7$

B. $5x = 9$

C. $x/3 - y/7 = 7$

D. $3/x + 5/y = 9$

E. $xy = 7$

F. $x^2 - 3y = 5$

NOTE
A, B, and C are lines. Coefficients may be negative or fractions or, as in B, B = 0. D is not a straight line since the letters are in the bottom. E is not a straight line since the variables are multiplied. F is not a straight line since the exponent of each letter must be 1.

DEFINITION
Slope—The slope (or slant) of a line is defined by $m = (y_2 - y_1)/(x_2 - x_1)$.

NOTES
1. The letter m is always used for the slope.

2. The 1s and 2s are subscripts, standing for point 1 and point 2. The 1 and the 2 mean x_1, x_2, y_1, and y_2 and stand for numbers, not variables, but I will not tell you what they are yet.

3. The calculus notation is $m = \Delta y/\Delta x$, where Δ is the Greek letter delta and means the change in y over the change in x.

4. The y's are always on top.

EXAMPLE 2, ON THE SLOPE—

We will graph the line that joins the points after this example and discuss further. Find the slope between

A. (2,3) and (6,12)

B. (4,−3) and (−1,3)

C. (1,3) and (6,3)

D. (2,5) and (2,8)

SOLUTIONS—

A. (2,3) and (6,12)

$x_1 y_1 \qquad x_2 y_2$

$$m = \frac{y_2 - y_1}{x_2 - x_1} = \frac{12 - 3}{6 - 2} = \frac{9}{4}$$

B. (4,−3) and (−1,3)

$x_1 \; y_1 \qquad x_2 \; y_2$

$$m = \frac{y_2 - y_1}{x_2 - x_1} = \frac{3 - (-3)}{-1 - 4} = \frac{6}{-5}$$

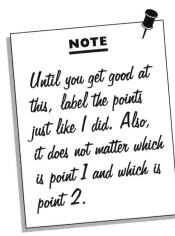

NOTE

Until you get good at this, label the points just like I did. Also, it does not matter which is point 1 and which is point 2.

NOTE

In 3 − (−3), the first minus sign is the one in the equation, and the second comes from the fact that y_1 is negative. *Be careful!*

C. (1,3) and (6,3)   m = (3 − 3)/6 − 1) = 0/5 = 0

D. (2,5) and (2,8)   m = (8 − 5)/(2 − 2) = 3/0
undefined, no slope, or infinite slope

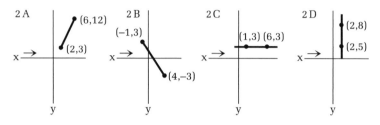

2A. If you walk in the direction of the arrow, when you get to the line and have to walk up, you always have a *positive slope.*

2B. If you have to walk down, you have a *negative slope.*

2C. Horizontal lines have m = 0.

2D. Vertical lines have no slope or infinite slope.

DEFINITION

Intercepts—x intercept, where the line hits the x-axis, is where y = 0. *y intercept* is where x = 0.

EXAMPLE 3

Given 3x − 4y = 12. Find the intercepts and graph the line.

x-intercept: y = 0. 3x = 12. x = 4. The point is (4,0). (x coordinate is always first.)

y-intercept: x = 0. −4y = 12. y = −3. The point is (0,−3), and the graph is as shown.

NOTE

This is the easiest way to graph a line with two intercepts.

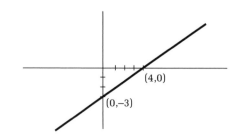

There are three exceptions, the ones with only one intercept:

x = 3 (all vertical lines are x = something).

y = 5 (all horizontal lines are y = something).

y = 3x [where the only intercept is (0,0)]. Pick some other point, say x = 2. So y = 6 and we get the second point (2,6).

Graphs are all included in the picture here.

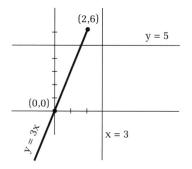

We are almost ready to find the equation of a line. There are four basic forms of a line. The two main ones are the point-slope and the slope-intercept.

DEFINITION

Slope-intercept—$y = mx + b$. When you solve for y, the coefficient of x is the slope. b is the y intercept—when $x = 0$, $y = b$. This is the most common form in high school books.

DEFINITION

Point-slope—$m = (y - y_1)/(x - x_1)$. You are given the slope and a point (x_1, y_1). The point-slope is given in different forms in different books. I will try to convince you that this form is the best to use because it eliminates many of the arithmetic fractions, which tend to bother too many of you.

We will now do some problems of this type. Although they are not long problems, most students have difficulty at first. Don't get discouraged if you do.

EXAMPLE 4—

Find the equation of the line through the points (2,3) and (7,11) and write the answer in standard form.

We will do this problem two different ways. I think you may be convinced the point-slope method is better.

Point-slope **(2,3) and (7,11)**	**Slope-intercept** **(2,3) and (7,11)**
$m = \dfrac{(11-3)}{(7-2)} = \dfrac{8}{5}$	$m = \dfrac{(11-3)}{(7-2)} = \dfrac{8}{5}$
$m = \dfrac{y - y_1}{x - x_1}$	$y = mx + b$
$\dfrac{8}{5} = \dfrac{y-3}{x-2}$	$y = \left(\dfrac{8}{5}\right)x + b$
$8(x-2) = 5(y-3)$	$3 = \left(\dfrac{8}{5}\right)(2) + b$
$8x - 16 = 5y - 15$	$3 = \left(\dfrac{16}{5}\right) + b$
$8x - 5y = 1$	$b = 3 - \dfrac{16}{5} = \dfrac{15}{5} - \dfrac{16}{5} = -\dfrac{1}{5}$
	$y = \left(\dfrac{8}{5}\right)x - \dfrac{1}{5}$
	$5y = 8x - 1$
	$-8x + 5y = -1 \text{ or } 8x - 5y = 1$

The reason the point-slope method is easier for most is that the only fractional skill is cross multiplication,[*] a skill most math students do very well. The slope-intercept method requires a number of fractional skills. About 90 percent of the time, point-slope is better.

At this stage of your mathematics, all of you should put the lines in standard form to practice your algebra skills. However, to save space and keep the book friendlier, I will leave the answer in either of these two forms.

EXAMPLE 5—

Write the equation of the line with slope 3, y intercept 7.

$y = mx + b$ is the easiest. $y = 3x + 7$.

EXAMPLE 6—

Write the equation of the line with slope -4 and x intercept 9.

Tricky. $m = -4$ and point (9,0). Use point-slope.

$$m = \frac{y - y_1}{x - x_1} \qquad -4 = \frac{y - 0}{x - 9}$$

EXAMPLE 7, PART A—

Find the equation of the line parallel to $3x + 4y = 5$ through (6,7).

EXAMPLE 7, PART B—

Find the equation of the line perpendicular to $3x + 4y = 5$ through (8,−9).

Parallel lines have the same slope. Perpendicular lines have negative reciprocal slopes. In either case we must solve for y.

[*]See the third-to-last line in the point-slope column.

$3x + 4y = 5$. $4y = -3x + 5$. $y = (-3/4)x + 5/4$.

We only care about the coefficient of x, which is the slope. The slope is $-3/4$.

A. *Parallel* means equal slopes. $m = -3/4$; the point is (6,7). The equation is

$$\frac{-3}{4} = \frac{y - 7}{x - 6}$$

B. *Perpendicular* means the slope is the negative reciprocal. $m = +4/3$; the point is (8,−9). The equation is

$$\frac{4}{3} = \frac{y - (-9)}{x - 8} \text{ or } \frac{4}{3} = \frac{y + 9}{x - 8}$$

PARABOLA

Later we will do a more complete study of the parabola. We need the basic, standard high school parabola for now. We will study parabolas of the form $y = ax^2 + bx + c$, $a \neq 0$. The coefficient of x^2 determines the parabola's shape.

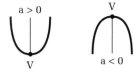

The low point (or high point), indicated by the letter V, is the vertex, which is why the letter V is usually used.

The x coordinate of the vertex is found by setting x equal to $-b/(2a)$. The y value is gotten by putting the x value into the equation for the parabola. The line through the vertex, the axis of symmetry, is given by $x = -b/(2a)$. The intercepts plus the vertex usually are enough for a fairly good picture. We will do some now.

EXAMPLE 8

$y = 2x^2 - 7x + 3$

Vertex $x = \dfrac{-b}{(2a)} = \dfrac{-(-7)}{2(2)} = \dfrac{7}{4}$

$y = 2\left(\dfrac{7}{4}\right)^2 - 7\left(\dfrac{7}{4}\right) + 3 = \dfrac{-25}{8}$ $\left(\dfrac{7}{4}, \dfrac{-25}{8}\right)$

Axis of symmetry: $x = \dfrac{-b}{(2a)}$ $x = \dfrac{7}{4}$

y intercept: $x = 0$, $y = 3$ $(0,3)$

x intercepts: $y = 0$

$2x^2 - 7x + 3 = (2x - 1)(x - 3) = 0$

$x = 1/2$, 3 and the intercepts are $(1/2,0)$ and $(3,0)$. The graph opens upward.

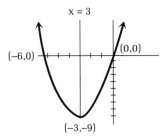

EXAMPLE 9

$y = x^2 + 6x$

Vertex $x = \dfrac{-b}{(2a)} = \dfrac{-6}{2}(1) = -3$

$y = (-3)^2 + 6(-3) = -9$ $(-3,-9)$

x intercepts: $y = 0$

$x^2 + 6x = x(x + 6) = 0$

$x = 0, -6(0, 0)$ and $(-6,0)$ are the x intercepts with $(0,0)$ also the y intercept. Picture is up. Axis of symmetry: $x = -3$.

EXAMPLE 10

$y = 9 - x^2$

Vertex $x = \dfrac{-b}{(2a)} = \dfrac{0}{2}(-1) = 0$

$y = 9$ $(0,-9)$ also y intercept

x intercepts: $y = 0$

$0 = 9 - x^2 = (3 - x)(3 + x)$

$x = 3, -3$; intercepts are $(3,0)$ and $(-3,0)$. Picture is down. Axis of symmetry: $x = 0$.

EXAMPLE 11—

$y = x^2 - 2x + 5$

Vertex $x = -(-2)/2(1) = 1$

$y = 1^2 - 2(1) + 5$ $(1,4)$

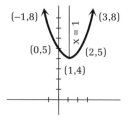

The y intercept is $(0,5)$. $y = x^2 - 2x + 5 = 0$. The quadratic formula gives imaginary roots, so no x intercepts.

To get more points, make a chart. Take two or three x values (integers) just below the vertex and two or three just above.

x	y
−1	8
0	5
1	4
2	5
3	8

Axis of symmetry: $x = 1$. Parabola is up. We are now ready for functions.

CHAPTER 14

DISTANCE FORMULA, MIDPOINT FORMULA, CIRCLE 11, AND ANALYTIC GEOMETRY PROOFS

DEFINITION

Distance—The distance between two points (x_1, y_1) and (x_2, y_2) is given by the formula
$d = \sqrt{(x_2 - x_1)^2 + (y_2 - y_1)^2}$.

DEFINITION

Midpoint—The average (mean) between two points is $[(x_1 + x_2)/2, (y_1 + y_2)/2]$.

EXAMPLE 1

Given $(3, -5)$ and $(9, -1)$. Find the length of the line segment between the points and the midpoint of that line segment.

$$d = \sqrt{\{(9 - 3)^2 + [-1 - (-5)]^2\}} = \sqrt{(36 + 16)} = \sqrt{52}$$
$$= \sqrt{2 \cdot 2 \cdot 13} = 2\sqrt{13}$$

Midpoint is $\{(3 + 9)/2, [-1 + (-5)]/2\} = (6, -3)$.

DEFINITION

Circle—Set of all points (x, y) at a distance r (don't tell anyone, but r stands for radius) from a given point (h, k) called the center of the circle.

The equation of a circle is the square of the distance formula $(x - h)^2 + (y - k)^2 = r^2$.

EXAMPLE 2—

Find the center and radius of $(x - 3)^2 + (y + 5)^2 = 7$.

C(3,–5) which is the opposite sign, and the radius = $\sqrt{7}$.

EXAMPLE 3—

Find r and C for the circle $2x^2 + 2y^2 - 12x + 6y + 8 = 0$.

First, we know it's a circle, since the coefficients of x^2 and y^2 are the same, as long as the radius turns out to be a positive number.

Coefficients of x^2 and y^2 must be 1.

$2x^2 + 2y^2 - 12x + 6y + 8 = 0$

or

Group x's and y's together, constant to other side.

$x^2 + y^2 - 6x + 3y + 4 = 0$

or

$x^2 - 6x + y^2 + 3y = -4$

or

Complete the square and add the term(s) to each side.

$x^2 - 6x + (-6/2)^2 + y^2 + 3y + (3/2)^2 = (-6/2)^2 + (3/2)^2 - 4$

or

Factor and do the arithmetic.

$(x - 3)^2 + (y + 3/2)^2 = 29/4$

The center is (3,–3/2) and the radius is $\sqrt{\dfrac{29}{4}} = \dfrac{\sqrt{29}}{2}$.

ANALYTIC GEOMETRY PROOFS

Analytic geometry, or algebraic geometry, used to be part of all geometry books. I feel compelled to show you two examples of what was and hopefully will be again in your books.

The picture we draw will be used for both proofs.

Theorem I   Prove the diagonals of a parallelogram bisect each other.

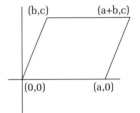

We must draw a parallelogram with the x and y axes (plural of axis) the easiest way possible. One vertex is at the origin, (0,0). Another vertex is on the x axis, point (a,0), length of side is "a." A third vertex is (b,c). The fourth vertex: the side is parallel to the one on the x axis and the same length. Height is the same and the x value is "a" units more. The point is (a + b,c). (This part takes practice.)

To show the diagonals bisect each other, we can show both diagonals have the same midpoint!!!! The midpoint between (0,0) and (a + b,c) = (a + b/2,c + 0/2). The midpoint between (a,0) and (b,c) = (a + b/2,c + 0/2). Since they are the same, the diagonals bisect each other. The proof itself is easy if (a big if) you set it up and see what has to be done.

Sometimes you are asked to do an analytic geometry proof for a numerical problem. Let's do one.

EXAMPLE 4—

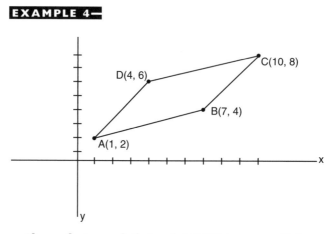

Show that quadrilateral ABCD is a parallelogram.

Show that the diagonals bisect each other.

1. There are two ways to show that a figure is a parallelogram:

   A. Show that opposite sides are equal (congruent), a theorem.

   B. Show that opposite sides are parallel (definition of a parallelogram).

   A. $\overline{AB} = \sqrt{(7-1)^2 + (4-2)^2} = \sqrt{40}$.

   $\overline{CD} = \sqrt{(10-4)^2 + (8-6)^2} = \sqrt{40}$.

   $\overline{AD} = \sqrt{(4-1)^2 + (6-2)^2} = 5$.

   $\overline{BC} = \sqrt{(10-7)^2 + (8-4)^2} = 5$.

   Since the opposite sides of the quadrilateral are the same, the figure is a parallelogram.

B. $m_{\overline{AB}} = \dfrac{4-2}{7-1} = \dfrac{1}{3}$; $m_{\overline{CD}} = \dfrac{8-6}{10-4} = \dfrac{1}{3}$;

$m_{\overline{AD}} = \dfrac{6-2}{4-1} = \dfrac{4}{3}$; $m_{\overline{BC}} = \dfrac{8-4}{10-7} = \dfrac{4}{3}$.

Since the opposite sides are parallel, the quadrilateral is a parallelogram.

Finding the **slopes** is almost always the easier way!

2. To show that the diagonals bisect each other, we will find the midpoint of each diagonal. If they are the same, then the diagonals really do bisect each other:

Midpoint $\overline{BD} = \left(\dfrac{4+7}{2}, \dfrac{6+4}{2} \right) = \left(\dfrac{11}{2}, 5 \right)$.

Midpoint $\overline{AC} = \left(\dfrac{10+1}{2}, \dfrac{8+2}{2} \right) = \left(\dfrac{11}{2}, 5 \right)$.

We have shown that the diagonals bisect each other.

Let's do one more proof.

Theorem 2 If the diagonals of a parallelogram are equal (congruent), the figure is a rectangle.

Lengths the same in algebra mean distance formula. We will use the *square* of distance, since if we don't, the first step will be to square both sides. Then we will simplify and see what happens.

Diagonals the same means

$(x_2 - x_1)^2 + (y_2 - y_1)^2 = (x_4 - x_3)^2 + (y_4 - y_3)^2$

$[(a+b) - 0]^2 + (c-0)^2 = (b-a)^2 + (c-0)^2$

Multiplying out, we get

$a^2 + 2ab + b^2 + c^2 = b^2 - 2ab + a^2 + c^2$

All square terms cancel. Sooo we get

$2ab = -2ab$ or $4ab = 0$.

Dividing by 4 we get that ab = 0. This means either a = 0 or b = 0.

Looking at the picture, *a* could not be 0. There would be no parallelogram. So the length b = 0. But if b = 0, b must be on the y axis, and the parallelogram is a rectangle!!!

These problems are becoming very popular. Let's do one more.

Given the isosceles trapezoid WXYZ.

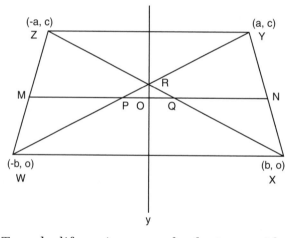

To make life easier, we make the trapezoid symmetric with respect to the y-axis.

M and N are the midpoints of sides $\overline{WZ}$ and $\overline{XY}$, respectively. Line segment $\overline{MN}$ is called the **median** of a trapezoid, which is a line segment connecting the midpoint of the two nonparallel sides of a trapezoid.

Before we start, let's make a few notes.

NOTE 1

The length of the top base, $\overline{ZY}$, is $a - (-a) = 2a$.

NOTE 2

The length of the bottom base, $\overline{WX} = b - (-b) = 2b$.

NOTE 3

The diagonals of the isosceles trapezoid are equal (congruent).

In this case, the length of $\overline{WY}$ and $\overline{XZ}$ is $\sqrt{(a + b)^2 + c^2}$.

Now for the questions:

1. Find M and N.

2. Show that $\overline{MN}$ is parallel to both bases.

3. Show that the length of $\overline{MN}$ equals half the sum of the bases.

4. Find the coordinates of P and Q.

5. Show that the length of $\overline{PQ}$ equals half the difference of the bases.

6. Show that the coordinates of point R are $[0, bc/(a + b)]$.

7. Show that the ratio of $\overline{RX}/\overline{ZX} = b / (a + b)$.

Now for the answers:

1. M, the midpoint of $\overline{WZ} = \left(\dfrac{(-a) + (-b)}{2}, \dfrac{c - 0}{2} \right)$

$= \left(\dfrac{-a - b}{2}, \dfrac{c}{2} \right)$.

N, the midpoint of $\overline{XY} = \left(\dfrac{a + b}{2}, \dfrac{c}{2} \right)$.

2. $m_{\overline{YZ}} = \dfrac{c - c}{a - (-a)} = \dfrac{0}{2a} = 0.$

$m_{\overline{WX}} = \dfrac{0 - 0}{b - (-b)} = 0.$

$m_{\overline{MN}} = \dfrac{\dfrac{c}{2} - \dfrac{c}{2}}{\dfrac{a + b}{2} - \dfrac{(-a) + (-b)}{2}} = \dfrac{0}{a + b} = 0.$

3. Half the sum of the bases is $(2a + 2b)/2 =$

   $2(a + b)/2 = a + b.$ Since the y values are the

   same, $\overline{MN} = (a + b)/2 - [(-a) + (-b)]/2 =$

   $a/2 + b/2 + a/2 + b/2 = a + b.$

4. P, the midpoint of $\overline{WY} = \left(\dfrac{a - b}{2}, \dfrac{c}{2}\right).$ Q, the mid-

   point of $\overline{XZ} = \left(\dfrac{-a + b}{2}, \dfrac{c}{2}\right).$

5. Half the difference of the bases is $(2b - 2a)/2 =$

   $2(b - a)/2 = b - a.$ The length of $\overline{PQ} = \dfrac{-a + b}{2} -$

   $\dfrac{a - b}{2} = \dfrac{b}{2} - \dfrac{a}{2} + \dfrac{b}{2} - \dfrac{a}{2} = b - a.$

6. We need to find the equation of the line through

   X and Z in point-slope form: $m = \dfrac{y - y_1}{x - x_1}.$

   $m = \dfrac{0 - c}{b - (-a)} = \dfrac{-c}{a + b}.$ $(x_1, y_1) = (b, 0).$

Sooo $\dfrac{-c}{a+b} = \dfrac{y-0}{x-b}$. $y = \left(\dfrac{-c}{a+b}\right)(x-b)$. For the

point R = (x, y), from the picture, the x value is 0.

$$y = \dfrac{bc}{a+b}.$$

7. $\dfrac{\overline{RX}}{\overline{ZX}} = \sqrt{\dfrac{(0-b)^2 + \left(\dfrac{bc}{a+b} - 0\right)^2}{[b-(-a)]^2 + (c-0)^2}} =$

$$\sqrt{\dfrac{\dfrac{(a+b)^2\,b^2 + b^2 c^2}{(a+b)^2}}{(a+b)^2 + c^2}} = \sqrt{\dfrac{\dfrac{b^2\,[(a+b)^2 + c^2]}{(a+b)^2}}{(a+b)^2 + c^2}}$$

$$\sqrt{\dfrac{b^2}{(a+b)^2}} = \dfrac{b}{a+b}.$$

NOTE

$$\dfrac{\overline{RZ}}{\overline{ZX}} = \dfrac{a}{a+b}$$

Whew! I certainly hope the teacher won't give you a problem with seven parts. But this is a very good problem to learn from.

These are neat proofs, not that easy and not like the proofs before. They improve your brain.

CHAPTER 15

FUNCTIONS, TRANSLATIONS, STRETCHES, CONTRACTIONS, AND FLIPS

One of the most neglected topics in high school is the study of functions. In this book there are three rather lengthy chapters directly related to functions and several others that are indirectly related. There are two reasons for this: functions are important, and most calculus courses assume you know this topic almost perfectly, an unrealistic assumption. So let's get started at the beginning.

Function: Given a set D. To each element in D, we assign one and only one element.

EXAMPLE I—

Does the picture here represent a function? The answer is yes. 1 goes into a, 2 goes into 3, 3 goes into 3, and 4 goes into *pig.* Each element in D is assigned one and only one element.

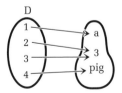

The next example will show what is *not* a function. But let us talk a little more about this example. The set D is called the *domain.* We usually think about x values when we think about the domain. This is not necessarily true, but it is true in nearly all high school and college courses, so we will assume it.

There is a second set that arises. It is not part of the definition. However, it is always there. It is called the *range.* Notice that the domain and the range can contain the same thing (the number 3) or vastly different things (3 and *pig*). However, in math, we deal mostly with numbers and letters. The rule (the arrows) is called the *map* or *mapping.* 1 is mapped into a; 2 is mapped into 3; 3 is mapped into 3; and 4 is mapped into *pig.*

FUNCTIONAL NOTATION

The rule is usually given in a different form: f(1) = a (read "f of 1 equals a"); f(2) = 3; f(3) = 3; and f(4) = *pig.*

NOTE I
When we think of the range, we will think of the y values, although again this is not necessarily true.

NOTE 2
We cannot always draw pictures of functions, and we will give more realistic examples after we give an example of something that is not a function.

EXAMPLE 2

The picture here does not represent a function, since 1 is assigned two values, a and d.

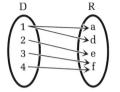

EXAMPLE 3

Let $f(x) = x^2 + 4x + 7$. D = {1,–3,10}.

$f(1) = (1)^2 + 4(1) + 7 = 12$

$f(-3) = (-3)^2 + 4(-3) + 7 = 4$

$f(10) = (10)^2 + 4(10) + 7 = 147$

The range would be {4,12,147}. If we graphed these points, we would graph (1,12), (–3,4), and (10,147).

NOTE
Instead of graphing points (x,y), we are graphing points (x,f(x)). For our purposes, the notation is different, but the meanings are the same.

EXAMPLE 4—

Let $g(x) = x^2 - 5x - 9$. $D = \{4,0,-3,a^4,x + h\}$. Find the elements in the range.

This is a pretty crazy example, but there are reasons to do it.

$$g(4) = (4)^2 - 5(4) - 9 = -13$$

$$g(0) = 0^2 - 5(0) - 9 = -9$$

$$g(-3) = (-3)^2 - 5(-3) - 9 = 15$$

$$g(a^4) = (a^4)^2 - 5a^4 - 9 = a^8 - 5a^4 - 9$$

$$g(x + h) = (x + h)^2 - 5(x + h) - 9$$

$$= x^2 + 2xh + h^2 - 5x - 5h - 9$$

The range is $\{-13,-9,15,a^8 - 5a^4 - 9,x^2 + 2xh + h^2 - 5x - 5h - 9\}$.

TRANSLATIONS, STRETCHES, CONTRACTIONS, FLIPS

If there was ever a more useless topic than this, I want to know it. It occurs in either this course or algebra 2. It not only wastes time, replacing something, anything, more valuable, it also can hurt you because when you sketch sin x or cos x in trig, the translating, stretching, contracting, and flipping is done much differently. But because it is in your book, here goes.

Suppose f(x) looks like this.

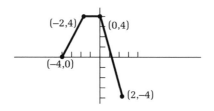

EXAMPLE 5—

f(x) + 3 would be 3 units up in the y direction and would look like this.

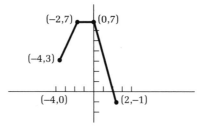

EXAMPLE 6—

f(x) − 4 is 4 units down. (Look at f(x), the original, first.)

EXAMPLE 7

2f(x): twice as high (or twice as low). (Points on the x-axis stay the same.)

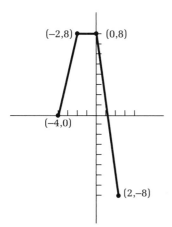

EXAMPLE 8

(1/4)f(x): 1/4 as high or as low.

EXAMPLE 9—

−f(x): upside down. Flip the x-axis. Top becomes bottom and bottom becomes top.

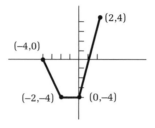

EXAMPLE 10—

f(x − 3): three units to the right.

EXAMPLE 11—

f(x + 5): five units to the left.

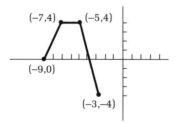

EXAMPLE 12—

f(−x): flip the y-axis. Right becomes left and left
becomes right.

EXAMPLE 13—

f(2x): compact, twice as close. (Points in the y-axis stay
the same.)

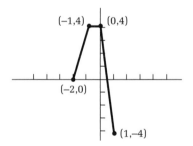

EXAMPLE 14—

$f\left(\dfrac{1}{3}x\right)$: stretch three times.

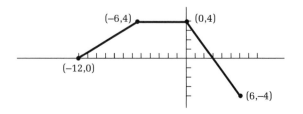

There can be combos: $y = -2f(x - 5) + 3$. Order of operations:

1. Inside parenthesis, five units to the right.

2. Multiplying the function −2: twice as high, upside down.

3. Up or down +3 up 3. Let's draw it in stages.

CHAPTER 16

PARABOLAS II, ELLIPSES, AND HYPERBOLAS

We will now discuss these three curves the way they are discussed in some precalc and algebra 2 books.

The important way to study these curves is to relate the equation to the picture. If you do this, this entire chapter will become much easier.

DEFINITION

Parabola—The set of all points equidistant from a point, called a *focus,* and a line, called the *directrix.* F is the focus. The point V, the *vertex,* is the closest point to the directrix.

NOTE

According to the definition $\overline{FV} = \overline{VR}$, $\overline{FP_1} = \overline{P_1R_1}$, $\overline{FP_2} = \overline{P_2R_2}$, etc.

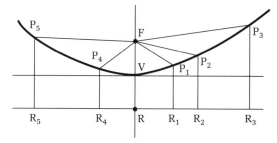

Let's do this development algebraically.

Let the vertex V be at (0,0) and the focus F be at (0,c). The directrix is at y = −c. Let P(x,y) be any point on the parabola. The definition of the parabola says $\overline{FP} = \overline{PQ}$. Everything on $\overline{PQ}$ has the same x value, and everything on y = −c has the same y value. Point Q has to be (x,−c). On $\overline{PQ}$, since the x values are the same, the length of the line is y − (−c). Using the distance formula to get $\overline{PF}$ and setting the two segments squared equal to each other, we get . . . $(x − 0)^2 + (y − c)^2 = (y + c)^2$. Squaring, we get $x^2 + y^2 − 2cy + c^2 = y^2 + 2cy + c^2$. Simplifying, we get $x^2 = 4cy$.

Here is a chart that will be helpful in relating the equation to the picture.

Vertex	Focus	Directrix	Equation	Picture	Comment
(0,0)	(0,c)	y = −c	$x^2 = 4cy$		The original derivation
(0,0)	(0,−c)	y = c	$x^2 = −4cy$		y replaced by −y
(0,0)	(c,0)	x = −c	$y^2 = 4cx$		x,y interchange in top line
(0,0)	(−c,0)	x = c	$y^2 = −4cx$		x replaced by −x in the above line

EXAMPLE 1

$y^2 = −7x$. Sketch; label vertex, focus, directrix.

The chart tells us the picture is the last line. Now let 4c equal 7, ignoring the minus sign. c = 7/4. Vertex is (0,0). Focus is (−7/4,0) because it is on the negative x axis. The directrix is x = +7/4, positive because it is to the right and x = 7/4 since it is a vertical line. Easy, isn't it?

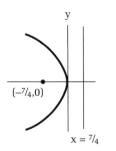

EXAMPLE 2

Sketch $(y − 3)^2 = −7(x + 2)$.

To understand this curve, we look at the difference between $x^2 + y^2 = 25$ and $(x - 3)^2 + (y + 6)^2 = 25$. Has the shape changed? No. Has the radius changed? No. The only thing that has changed is its position. The center is now at the point $(3,-6)$ instead of at the point $(0,0)$.

In the case of our little parabola, it is the vertex that has changed. $V = (-2,3)$. (Remember x is always first.) 4c is still equal to 7. $c = 7/4$, but F is $(-2 - 7/4,3)$, 7/4 to the left of the vertex. The directrix is $x = -2 + 7/4$. I do not do the arithmetic here so that you know where the numbers come from.

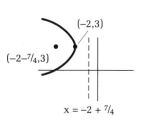

EXAMPLE 3—

Sketch the parabola $2x^2 + 8x + 6y + 10 = 0$.

$2x^2 + 8x + 6y + 10 = 0$	**Divide by coefficient of x^2 or y^2.**
$x^2 + 4x + 3y + 5 = 0$	**On the left get all the terms related to the square term, everything else on the other side.**
$x^2 + 4x = -3y - 5$	**Complete the square; add to both sides.**
$x^2 + 4x + 4 = -3y - 5 + 4$	**Factor; do the arithmetic.**
$(x + 2)^2 = -3y - 1$	**Weird last step. No matter what the coefficient on the right, factor it all out, even if it results in a fraction in the parentheses.**

$$(x + 2)^2 = -3\left(y + \frac{1}{3}\right)$$

Sketch; the picture is on the following page.
$V(-2,-1/3)$. $4c = 3$. $c = 3/4$. $F(-2,-1/3 - 3/4)$.
Directrix $y = -1/3 + 3/4$.

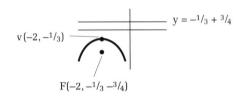

$y = -1/3 + 3/4$

$v(-2, -1/3)$

$F(-2, -1/3 - 3/4)$

DEFINITION

Algebraically, the *ellipse* is defined as $\overline{PF_1} + \overline{PF_2} = 2a$. $2a > 2c$. $2c$ is the distance between the *foci*. a will be determined later. P is any point on the ellipse. To paraphrase the definition, given two points called foci, we draw a line from one of the points to any point on the curve and then from that point on the curve to the other focus; if the sum of the two lengths always adds to the same number, 2a, the figure formed will be an ellipse.

I know you'd all desperately like to draw an ellipse. Take a nonelastic string. Attach thumbtacks at either end. Take a pencil and stretch the string as far as it goes. Go 360°. You will trace an ellipse.

Some of you have seen the equation of an ellipse, but few of you have seen its derivation. It is an excellent algebraic exercise for you to try. You will see there is a lot of algebra behind a very simple equation.

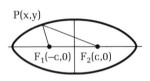

P(x,y)

$F_1(-c,0)$ $F_2(c,0)$

Use the distance formula.	$\overline{PF_1} + \overline{PF_2} = 2a$
Isolate one square root and square both sides.	$\sqrt{(x - (-c))^2 + (y - 0)^2}$ $+ \sqrt{(x - c)^2 + (y - 0)^2} = 2a$
Do the algebra.	$[\sqrt{(x + c)^2 + y^2}]^2 = [2a - \sqrt{(x - c)^2 + y^2}]^2$
Combine like terms; isolate the radical.	$x^2 + 2cx + c^2 + y^2 = 4a^2 + x^2 - 2cx + c^2 + y^2$ $- 4a\sqrt{(x - c)^2 + y^2}$
Divide by 4; again square both sides.	$4a\sqrt{(x - c)^2 + y^2} = 4a^2 - 4cx$
Do the algebra.	$[a\sqrt{(x - c)^2 + y^2}]^2 = (4a^2 - 4cx)^2$

$$a^2(x^2 - 2cx + c^2 + y^2) = a^4 - 2a^2cx + c^2x^2$$

or $\quad a^2x^2 - c^2x^2 + a^2y^2 = a^4 - a^2c^2$

$$\frac{(a^2 - c^2)x^2}{(a^2 - c^2)a^2} + \frac{a^2y^2}{a^2(a^2 - c^2)} = \frac{a^2(a^2 - c^2)}{a^2(a^2 - c^2)}$$

Factor out x^2 on the left two terms and a^2 from the right two terms; divide by $(a^2 - c^2) \times a^2$.

We get

$$\frac{x^2}{a^2} + \frac{y^2}{a^2 - c^2} = 1 \qquad \text{or} \qquad \frac{x^2}{a^2} + \frac{y^2}{b^2} = 1$$

letting $a^2 - c^2 = b^2$. Whew!!!!!!

We are still not finished. Let's find out what a is and what b is. $\overline{F_1P} + \overline{PF_2} = 2a$, where P is any point on the ellipse. Let T be the point P. $\overline{F_1T} + \overline{TF_2} = 2a$. By symmetry, $\overline{F_1T} = \overline{TF_2}$. So $\overline{F_1T} = \overline{TF_2} = a$. Since $a^2 - c^2 = b^2$, $\overline{OT} = \overline{OT'} = b$. The coordinates of T are $(0,b)$; T′ is $(0,-b)$.

We would like to find the coordinates of U (and U′), but the letters a, b, and c are used up. Oh well, let's see what happens. $\overline{F_2U} + \overline{UF_1} = 2a$. $\overline{F_2U} = x - c$. $\overline{UF_1} = x + c$. $x - c + x + c = 2a$. So $2x = 2a$; $x = a$. U is $(a,0)$; U′ is $(-a,0)$.

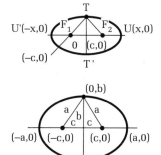

c = half the distance between the foci

b = length of the semiminor axis ("semi" means half; "minor" means smaller; "axis" means line)

a = length of semimajor axis = distance from a focus to a minor vertex

$(\pm a,0)$—major vertices

$(0,\pm b)$—minor vertices

$(\pm c,0)$—foci, always located on the major axis

Although the derivation was long, sketching should be short.

EXAMPLE 4—

Sketch $x^2/7 + y^2/5 = 1$.

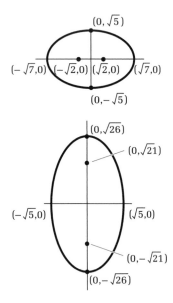

In the case of the ellipse, the longer axis is indicated by the larger number under x^2 or y^2. Try not to remember a or b. Remember the picture! This is longer in the x direction. $y = 0$, major vertices $(\pm\sqrt{7},0)$; $x = 0$, minor vertices $(0,\pm\sqrt{5})$. $c = \sqrt{(7 - 5)}$. The foci are $(\pm\sqrt{2},0)$.

EXAMPLE 5—

$$\frac{x^2}{5} + \frac{y^2}{26} = 1$$

Longer in the y direction. Major vertices: $(0,\pm26^{1/2})$. Minor vertices: $(0,\pm\sqrt{5})$. $c = \sqrt{(26 - 5)}$. Foci: $(0,\pm\sqrt{21})$. Foci are always on the longer axis.

EXAMPLE 6—

$$\frac{(x - 6)^2}{7} + \frac{(y + 4)^2}{5} = 1$$

This is the same basic example as Example 4, except the middle is no longer at $(0,0)$. It is at the point $(6,-4)$. Major vertices: $(6 \pm \sqrt{7},-4)$. Minor vertices: $(6,-4 \pm \sqrt{5})$. Foci: $(6 \pm \sqrt{2},-4)$.

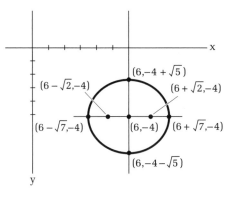

NOTE

The numbers found in Example 4 are added and subtracted from the appropriate coordinate of the center

(−6,−4). Also note that weird numbers were purposely chosen so that you could see where the numbers came from.

EXAMPLE 6 REVISITED—

$$\frac{(x - 6)^2}{7} + \frac{(y + 4)^2}{5} = 1$$

Just a few months after my last book went to press, I realized a second way to find the vertices that would make it much clearer for you. So now I finally have a chance to show you.

The center is again at the point (6,−4). The vertices are directly east-west or north-south of (6,−4). All points east-west of (6,−4) must have the same y value, y = −4. Soooooo

$$\frac{(x - 6)^2}{7} + \frac{(-4 + 4)^2}{5} = 1$$

$$\frac{(x - 6)^2}{7} + 0 = 1$$

$$(x - 6)^2 = 7$$

$$x - 6 = \pm\sqrt{7}$$

Soooooo x = 6 ± $\sqrt{7}$ and the major vertices are (6 ± $\sqrt{7}$,−4). Points north-south of (6,−4) have the same x values, x = 6.

$$\frac{(6 - 6)^2}{7} + \frac{(y + 4)^2}{5} = 1$$

$$0 + \frac{(y + 4)^2}{5} = 1$$

$$(y + 4)^2 = 5$$

$$y + 4 = \pm\sqrt{5}$$

Therefore $y = -4 \pm \sqrt{5}$ and the minor vertices are $(6, -4 \pm \sqrt{5})$. For the same reasons, the foci, always on the major axis, are $(6 \pm \sqrt{2}, -4)$. The sketch is, of course, the same.

EXAMPLE 7—

Sketch and discuss $4x^2 + 5y^2 + 30y - 40x + 45 = 0$.

Like the parabola and circle, we must complete the square, only a little differently.

Group the x and y terms; number to the other side.

$$4x^2 + 5y^2 + 30y - 40x + 45 = 0$$

$$4x^2 - 40x + 5y^2 + 30y = -45$$

Factor out coefficients of x^2 and y^2; complete the square in the parentheses; add the number term inside the parentheses multiplied by the number outside the parentheses to each side—with both x and y. Then do the arithmetic and divide by 100 to get 1 on the right.

$$4\left[x^2 - 10x + \left(\frac{-10}{2}\right)^2\right] + 5\left[y^2 + 6y + \left(\frac{6}{2}\right)^2\right]$$

$$= -45 + 4\left(\frac{-10}{2}\right)^2 + 5\left(\frac{6}{2}\right)^2$$

$$\frac{4(x - 5)^2}{100} + \frac{5(y + 3)^2}{100} = \frac{100}{100}$$

$$\text{or} \quad \frac{(x - 5)^2}{25} + \frac{(y + 3)^2}{20} = 1$$

The center is $(5, -3)$. Under the $(x - 5)^2$ term is larger; the ellipse is long in the x direction. $\sqrt{25}$ units are to the left and right of the center. Major vertices are $(5 \pm \sqrt{25}, -3)$.

Under the $(y + 3)^2$, $\sqrt{20}$ above and below the center. Minor vertices are $(5, -3 \pm \sqrt{20})$. $c = \sqrt{(25 - 20)}$. Foci, on the larger axis, are $(5 \pm \sqrt{5}, -3)$.

   Of course you should use 5 instead of $\sqrt{25}$, but I left $\sqrt{25}$ to show you where the 5 came from.

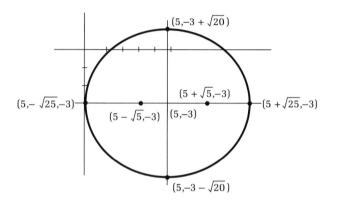

DEFINITION

Hyperbola—set of all points P such that $\overline{F_1P} - \overline{PF_2} = 2a$. The derivation is exactly the same as for the ellipse. Once is enough!!!!! The equation is $x^2/a^2 - y^2/b^2 = 1$, where $a^2 + b^2 = c^2$. $(\pm a, 0)$—transverse vertices. $(\pm c, 0)$—foci. Asymptotes: $y = \pm(b/a)x$. Slopes of the lines are the square root of the number under the y^2 term over the square root of the number under the x^2 term. The shape of the curve depends on the location of the minus sign, *not* the largeness of the numbers under the x^2 or y^2 term.

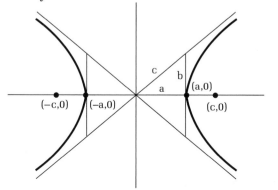

EXAMPLE 8—

Sketch and label $x^2/7 - y^2/11 = 1$.

Transverse vertices: $y = 0$ (set the letter after the minus sign equal to 0). $(\pm\sqrt{7},0)$. $c = \sqrt{(7 + 11)}$. Foci: $(\pm\sqrt{18},0)$. Asymptotes: $y = \pm(\sqrt{11}/\sqrt{7})x$.

NOTE

Curve does not hit y axis. If $x = 0$, $y = \pm(-11)^{1/2}$, which are imaginary.

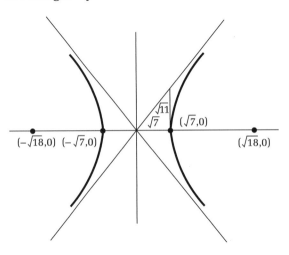

EXAMPLE 9—

Sketch and discuss $y^2/5 - x^2/9 = 1$.

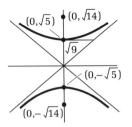

Let $x = 0$. Transverse vertices are $(0,\pm\sqrt{5})$. $c = \sqrt{(5 + 9)}$. Foci are $(0,\pm\sqrt{14})$. Asymptotes: $y = \pm(\sqrt{5}/\sqrt{9})x$. The sketch is as shown here.

EXAMPLE 10—

Sketch and discuss $(y - 6)^2/5 - (x + 7)^2/9 = 1$.

This is the same as Example 9, except the "center" of the hyperbola, where the asymptotes cross, is no longer at $(0,0)$. The center is $(-7,6)$. $(x + 7)^2 = 0$; then $a = \sqrt{5}$, $\sqrt{5}$ above and below the center. $c = \sqrt{(5 + 9)}$. The foci are $\sqrt{14}$ above and below the center. V($-7,6 \pm \sqrt{5}$). F($-7,6 \pm \sqrt{14}$). Asymptotes: $y - 6 = \pm(\sqrt{5}/\sqrt{9})(x + 7)$.

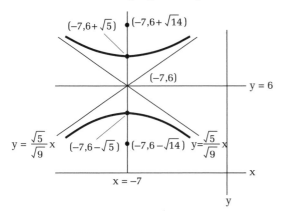

EXAMPLE 10 REVISITED—

$$\frac{(y - 6)^2}{5} - \frac{(x + 7)^2}{9} = 1$$

Just like with the ellipse, let's clarify the location of the vertices. In this case, the vertices occur north-south of the center $(-7,6)$. So the x value of the vertices is the same as the center, $(-7,6)$. $x = -7$. Substituting, we get

$$\frac{(y - 6)^2}{5} - \frac{(-7 + 7)^2}{9} = 1$$

$$(y - 6)^2 = 5$$

$$y = 6 \pm \sqrt{5}$$

and the vertices are $(-7,6 \pm \sqrt{5})$. For the same reason, the foci are $(-7,6 \pm \sqrt{14})$. The asymptotes and the sketch are the same!

EXAMPLE 11—

Sketch and discuss $25x^2 - 4y^2 + 50x - 12y + 116 = 0$.

For the last time, we will complete the square, again a little differently than the other times. We will use exactly the same steps as for the ellipse, except for the minus sign.

$$25x^2 - 4y^2 + 50x - 12y + 116 = 0$$

$$25x^2 + 50x - 4y^2 - 12y = -116$$

$$25\left[x^2 + 2x + \left(\frac{2}{2}\right)^2\right] - 4\left[y^2 + 3y + \left(\frac{3}{2}\right)^2\right]$$

$$= -116 + 25\left(\frac{2}{2}\right)^2 - 4\left(\frac{3}{2}\right)^2$$

$$\frac{25(x+1)^2}{-100} - \frac{4\left(y+\dfrac{3}{2}\right)^2}{-100} = \frac{-100}{-100}$$

$$\frac{\left(y+\dfrac{3}{2}\right)^2}{25} - \frac{(x+1)^2}{4} = 1$$

The center is $(-1,-3/2)$. $V(-1,-3/2 \pm \sqrt{25})$. $F(-1,-3/2 \pm \sqrt{29})$. Asymptotes: $y + 3/2 = \pm(\sqrt{25}/\sqrt{4})(x+1)$.

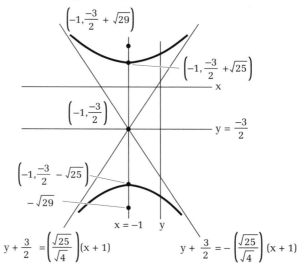

Sometimes we have a puzzle. Given some information, can we find the equation? You must always draw the picture and relate the picture to its equation.

EXAMPLE 12—

Find the equation of the parabola with focus (1,3), directrix x = 11.

Drawing F and the directrix, the picture must be the one shown here. The vertex is halfway between the x numbers. So x = (11 + 1)/2 = 6. V(6,3). c = the distance between V and F = 5. The equation is $(y - 3)^2 = -4c(x - 6) = -20(x - 6)$. Remember, the minus sign is from the shape and c is always positive for these problems.

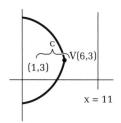

EXAMPLE 13—

Vertices are (2,3) and (12,3), and there is one focus (11,3). Find the equation of the ellipse.

The two vertices give the center [(12 + 2)/2,3] = (7,3). F(11,3). $(x - 7)^2/a^2 + (y - 3)^2/b^2 = 1$. a = 12 − 7 = 5. c = 11 − 7 = 4. $a^2 - b^2 = c^2$. $5^2 - b^2 = 4^2$. $b^2 = 9$ (no need for b). $(x - 7)^2/25 + (y - 3)^2/9 = 1$.

EXAMPLE 14—

Find the equation of the hyperbola with vertices (0,±6) and asymptotes y = ±(3/2)x.

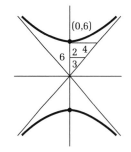

V(0,±6) says the center is (0,0) and the shape is $y^2/36 - x^2/a^2 = 1$. The slope of the asymptotes is 3/2 = square root of the number under y^2 over the square root of the number under the x^2 term. So 3/2 = 6/a. So a = 4. So $a^2 = 4^2 = 16$. The equation is $y^2/36 - x^2/16 = 1$.

This kind of question is shorter in length, but it does take practice. So practice!!!

CHAPTER 17

RIGHT AND NOT SO RIGHT-ANGLE TRIG, LAW OF SINES, AND LAW OF COSINES

Most geometry courses have a little right-angle trig. I am doing it the standard way, because the way I like to do it is too long for this book. If you want to see a better way, go to my *Precalc with Trig for the Clueless.* The trig part of the book is something you all will be able to understand. The rest depends on your algebra background.

We have a right triangle ABC, with the right angle at C. For angle A, $\overline{AC}$ is called the *adjacent leg* because it touches angle A. For angle A, $\overline{BC}$ is the *opposite leg* because it doesn't touch angle A. $\overline{AB}$ is of course the *hypotenuse.*

DEFINITION

$$Sine\ A = \sin A = \frac{\overline{BC}}{\overline{AB}} = \frac{\text{opposite}}{\text{hypotenuse}}$$

$$Cosine\ A = \cos A = \frac{\overline{AC}}{\overline{AB}} = \frac{\text{adjacent}}{\text{hypotenuse}}$$

$$Tangent\ A = \tan A = \frac{\overline{BC}}{\overline{AC}} = \frac{\text{opposite}}{\text{adjacent}}$$

Those are called *trig ratios.*

NOTE 1
If you extend or shrink the sides and if the angle stays the same, then the ratio of the sides stays the same because of similar triangles. Sooo the sine, cosine, and tangent of the angle will always remain the same.

NOTE 2
To find these values, we either need tables or a calculator. We will use calculators because that is what calculators should be used for.

NOTE 3
For angle B, the opposite side for A is the adjacent side for B, and the adjacent side for A is the opposite side for B. The hypotenuse is always the hypotenuse.

NOTE 4
There is a memory device to help remember these trig ratios, one that my father remembered from 1916 (anything lasting that long must be good). It is the Indian chief SOHCAHTOA: Sine is opposite over hypotenuse; cosine is adjacent over hypotenuse; and tangent is opposite over adjacent. By the way there are Indian powwows in my area. I've asked them if they object to the term "Indian." None of them did. Only politically correct people seem to want Native Americans.

Let's do some examples.

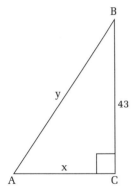

EXAMPLE 1

$\overline{BC}$ is 43 feet. Angle A is 73 degrees. Find all the missing parts.

1. Angle B = $90° - 73° = 17°$.

2. Tan $73°$ = opp/adj = $43/x$. Sooo x = $43/\tan 73°$. By calculator x = 13.1 feet.

3. Sin 73° = opp/hyp = 43/*y*. *y* = 43/sin 73°. By calculator y = 45.0 feet. Pretty easy huh?!!! Let's try another.

EXAMPLE 2—

Find x, angles A and B.

1. To get x, use old Pythagoras $x = \sqrt{87^2 + 68^2}$. By calculator x = 110.4 feet.

2. Tan A = opp/adj = 68/87. To find the angle, you must use $\tan^{-1}$ 68/85 = 38.7° = (m)∠A. (m)∠B = 90° − 38.7° = 51.3°.

NOTE

To know what $\tan^{-1}$, "arctan," or "inverse tan" is, you must study more trig. It really isn't hard or shouldn't be. At this point, all you need to know is that it gives you the angle.

Some books do problems on the angle of elevation and the angle of depression.

DEFINITION

Angle of elevation: The angle you look up at. The bottom line is the horizon.

> **IMPORTANT NOTE**
>
> *Make sure your calculator gives you degrees when you are doing angles.*

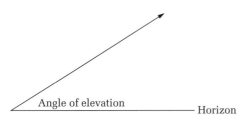

Angle of elevation —————— Horizon

DEFINITION

Angle of depression: The angle you look down at something. The top line is the horizon.

EXAMPLE 3—

Sandy wants to measure the height of the school. If the angle of elevation from the ground is 62 degrees, and Sandy is 20 feet from the school, how tall is the school?

$$\tan 62° = \frac{\text{opp}}{\text{adj}} = \frac{x}{20} \qquad x = 20 \tan 62° = 37.6 \text{ feet tall!!}$$

You might actually try to measure the height of your school.

EXAMPLE 4—

A man is looking down from the top of a 120-foot lighthouse. The angle of depression to a boat is 20°. How far away is the boat?

Try to get the variable in the numerator. $\cot 20° = x/120$. So $x = 120 \cot 20°$. By calculator, the answer is 330 feet.

EXAMPLE 5—

John is 200 feet from a building on which is located a tall antenna. The angle of elevation to the bottom of the antenna is 70°. The angle of elevation to the top of the antenna is 78°. How tall is the antenna?

There are two unknowns: the height of the building and the height of the antenna. We always want to try to get the unknown we don't want (the building) in the numerator. It makes the problem easier. In this problem, both unknowns are in the numerator when you use tangent.

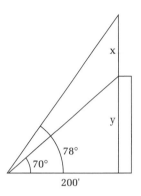

$$\tan 78° = \frac{(x + y)}{200} \qquad x + y = 200 \tan 78°$$

$$\tan 70° = \frac{y}{200} \qquad y = 200 \tan 70°$$

Subtract; factor.

$$x = 200(\tan 78° - \tan 70°) = 394 \text{ feet (big!)}$$

THE LAW OF SINES

When we don't have a right angle, we must use the law of sines or the law of cosines.

DEFINITION
Law of sines—a/sin A = b/sin B = c/sin C.

NOTE
For this exercise we will say sin 45° = 0.7, which is approximately correct.

We use the law of sines whenever we have any two angles and a side or two sides and an angle opposite one of those sides.

NOTE

Side is small letter; angle opposite is the same letter, only capitalized.

Two Angles and a Side

If we have two angles, we have the third angle. With a side, we learned, in geometry, that triangles are congruent by angle, side, angle. Therefore one solution is possible.

EXAMPLE 6—

A = 56°, B = 73°, and a = 20. Find all the other parts.

C = 180° − (56° + 73°) = 51°   a/sin A = b/sin B
20/sin 56° = b/sin 73°   b = 20 sin 73°/sin 56°   b = 263
a/sin A = c/sin C   20/sin 56° = c/sin 51°   c = 20 sin
51°/sin 56°   c = 19

Two Sides and an Angle Opposite One of Those Sides

If you draw the triangle, we have side-side-angle. From geometry, we know triangles are *not* congruent. This is called the *ambiguous case*. No triangle, one triangle, or two triangles are possible. Let us give one example of each.

EXAMPLE 7—

Let a = 10, A = 30°, b = 50. a/sin A = b/sin B. 10/
sin 30° = 50/sin B. sin B = (50 sin 30°)/10 = 2.5. No
triangle, since the sine is never bigger than 1.

EXAMPLE 8—

A = 135°, a = 70, b = 50. a/sin A = b/sin B. 70/
sin 135° = 50/sin B. sin B = (50 sin 135°)/70. sin B = ½.
B = 30° or 180° − 30° = 150°.

   135° + 30° = 165° is OK. 135° + 150° = 285° is no
good—the sum of angles of a triangle is 180°. One triangle is possible. A = 135°, B = 30°, so C = 15°. a = 70,
b = 50. a/sin A = c/sin C. 70/sin 135° = c/sin 15°. c =
70 sin 15°/sin 135°. c = 26.

EXAMPLE 9—

C = 30°. c = 5. d = 7. c/sin C = d/sin D. 5/sin 30 = 7/
sin D. sin D = (7 sin 30°)/5 = 0.7. D = 45° or D′ =
180° − 45° = 135°. 30° + 45°, OK. 30° + 135°, OK.
Two triangles!!!

Triangle 1: C = 30°, D = 45°, E = 105°. c = 5, d = 7, e/sin 105° = 5/sin 30°. e = 9.7.

Triangle 2: C = 30°, D′ = 135°, E′ = 15°. c = 5, d = 7, e′/sin 15° = 5/sin 30°. e′ = 2.6.

EXAMPLE 10—

The angle between Zeb and Sam as seen by Don is 70°. The angle between Don and Zeb as seen by Sam is 62°. Zeb and Sam are 70 feet apart. How far apart are Don and Sam?

Angle at Zeb is 48°. z/sin 48° = 70/sin 70°. z = 55 feet.

THE LAW OF COSINES

We use the law of cosines whenever we have three sides or two sides and the included angle.

DEFINITION

Law of cosines—$c^2 = a^2 + b^2 - 2ab \cos C$. Again, note angle C opposite side c. In each case only one triangle is possible, since triangles are congruent by side-side-side or side-angle-side.

EXAMPLE 11—

a = 3, b = 5, c = 7. $c^2 = a^2 + b^2 - 2ab \cos C$. $7^2 = 3^2 + 5^2 - 2(3)(5) \cos C$. Careful of the arithmetic. $15 = -30 \cos C$. $\cos C = -\frac{1}{2}$. C = 120° (quadrant II).

NOTE

Whenever you have four parts of a triangle, it is easier to use the law of sines. 3/sin A = 7/sin 120°. A = 22°. B = 180° − (22° + 120°) = 38°.

EXAMPLE 12—

$x = 10$, $y = 20$, $Z = 40°$. $z^2 = x^2 + y^2 - 2xy \cos Z$. $z^2 =$ $10^2 + 20^2 - 2(10)(20) \cos 40°$. $z = 13.9$. $13.9/\sin 40° = 10/\sin X$. $\sin X = 10(\sin 40°)/13.9$. $X = \sin^{-1}(10 \sin 40°/13.9)$. $X = 28°$. $Y = 180° - (28° + 40°) = 112°$.

EXAMPLE 13—

A plane travels east for 200 miles. It turns at a 25° angle to the north for 130 miles. How many miles from home is the plane?

$x^2 = 200^2 + 130^2 - 2(200)(130) \cos 155°$

$x = 323$ miles

That's all the trig we need for now. More? See *Precalc with Trig for the Clueless.*

CHAPTER 18

CONSTRUCTIONS

After many years, constructions are finally returning to geometry. I think they are fun as well as educational. I have included the most common. As the course changes, more will be added in the future.

EXAMPLE 1—

Bisect a given line segment.

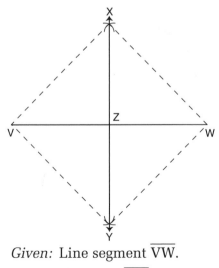

Given: Line segment $\overline{VW}$.

Required: Bisect $\overline{VW}$.

Construction

1. With V as the center and a radius more than $\frac{1}{2}$ the length of $\overline{VW}$, construct arcs above and below $\overline{VW}$.

2. With W as the center, using the same radius, construct similar arcs that intersect at X and Y.

3. Draw $\overline{XY}$, which bisects $\overline{VW}$ at Z. $\overline{VZ} \cong (=) \overline{ZW}$.

Paragraph Proof If $\overline{XV}$, $\overline{XW}$, $\overline{YV}$, and $\overline{YW}$, then the construct makes $\overline{XV} \cong (=) \overline{XW}$ and $\overline{YV} \cong (=) \overline{YW}$. This makes $\overline{XY}$ the perpendicular bisector of $\overline{VW}$ because V and W, two points, each equidistant from the endpoints of a line segment, form a perpendicular bisector.

EXAMPLE 2—

Bisect a given angle.

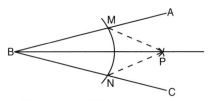

Given: ∠ABC.

Required: Bisect ∠ABC.

Construction

1. With B as the center and any radius (not too small), make an arc. The arc hits the angle at M and N.

2. With M and N as centers, with two reasonable congruent radii, draw two arcs that meet at P.

3. Draw $\overline{BP}$, which bisects ∠ABC. ∠ABP $\cong$ (=) ∠PBC.

Paragraph Proof If $\overline{MP}$ and $\overline{NP}$ are drawn, the construction makes these segments $\cong$ (=). Also, $\overline{BM} \cong$ (=) $\overline{BN}$. With $\overline{BP} \cong$ (=) $\overline{BP}$, $\triangle BMP \cong \triangle BNP$. By CPCTC (CPCTE), $\angle APF \cong$ (=) $\angle FPE$.

EXAMPLE 3

Construct a line perpendicular to a given line through a given point on the line.

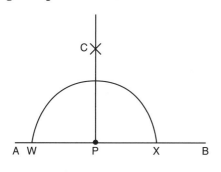

Given: Point P on line $\overset{\leftrightarrow}{AB}$.

Construct: a line through P $\perp \overset{\leftrightarrow}{AB}$.

Construction

1. With P as a center and any reasonable radius, draw an arc, that intersects $\overset{\leftrightarrow}{AB}$ at W and X.

2. With W and X as centers and with a radius bigger than $\overline{WP}$, draw arcs that intersect at point C.

3. Draw $\overline{PC}$, which is perpendicular to $\overset{\leftrightarrow}{AB}$ at P.

Paragraph Proof   The construction bisects straight angle APB. Congruent (equal) supplementary angles form right angles, which form perpendiculars.

EXAMPLE 4

Construct a line perpendicular to a given line from a point not on the line.

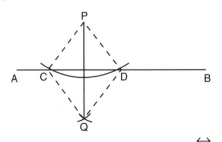

Given: Point P not on line $\overleftrightarrow{AB}$.

Construct: a line through $P \perp \overleftrightarrow{AB}$.

Construction

1. With P as center and any reasonable radius, draw an arc intersecting $\overleftrightarrow{AB}$ at C and D.

2. Using C and D as centers with equal radii greater than half of $\overline{CD}$, draw two arcs intersecting at point Q.

3. $\overline{PQ} \perp \overleftrightarrow{AB}$.

Paragraph Proof   If $\overline{PC}$, $\overline{PD}$, $\overline{QC}$, and $\overline{QD}$ are drawn, the construction makes $\overline{PC} \cong (=) \overline{PD}$ and $\overline{QC} \cong (=) \overline{QD}$. Therefore, $\overline{PQ}$ forms the perpendicular bisector of $\overline{CD}$ because P and Q are two points each equidistant from the endpoints of a line segment. Soooo, $\overline{PQ} \perp \overleftrightarrow{AB}$.

I hope you are having fun with these. These problems were my favorites. I liked the drawings. The math wasn't bad either. Let's do some more.

EXAMPLE 5—

Construct an angle congruent (equal) to another angle.

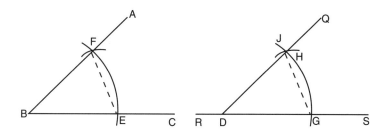

Given: ∠ABC and point D.

Construct: an angle at D the same as ∠ABC.

Construction

1. Through D, draw line $\overleftrightarrow{RS}$. $\overleftrightarrow{RS}$ doesn't have to be in the same direction as $\overline{BC}$, but it's prettier that way.

2. With B as a center, and any reasonable radius, draw an arc that intersects $\overrightarrow{BC}$ at E and $\overrightarrow{BA}$ at F.

3. With D as the center and using the same radius as in 2, draw an arc that intersects $\overrightarrow{DS}$ at G.

4. Measure $\overline{EF}$. With G as the center and $\overline{EF}$ as the radius, draw an arc, intersecting arc $\overset{\frown}{JG}$ at H.

5. Draw $\overleftrightarrow{DH}$. ∠HDY ≅ (=) ∠ABC.

Paragraph Proof $\overline{EF}$ and $\overline{GH}$ are drawn the same as each other. $\overline{BF}$, $\overline{BE}$, $\overline{DH}$, and $\overline{DG}$ are all the same. △BEF ≅ △DGH by SSS. So the two angles are congruent by corresponding parts.

EXAMPLE 6—

Construct a line parallel to another line.

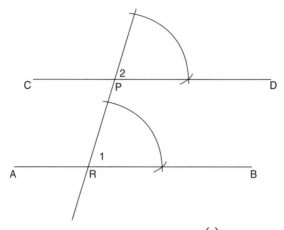

Given: Point P outside line $\overleftrightarrow{AB}$.

Construct: through P a line parallel to $\overleftrightarrow{AB}$.

Construction

1. Through P, draw a transversal, hitting $\overleftrightarrow{AB}$ at R.

2. At P, construct $\angle 1 \cong (=) \angle 2$ (as in example 5), a pair of corresponding angles.

3. Draw $\overleftrightarrow{CD}$ through point P.

Paragraph Proof   Since angle 1 and angle 2 are constructed the same, corresponding angles make the two lines parallel.

EXAMPLE 7—

Construct a tangent to a given circle through a given point on the circle.

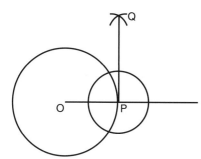

Given: Point P on circle O.

Construct: a tangent to the circle O at point P.

Construction

1. Draw the radius $\overline{OP}$ and extend it outside the circle.
2. At P construct the perpendicular $\overline{PQ} \perp \overline{OP}$ as in example 3.
3. $\overline{PQ}$ is the required tangent.

Paragraph Proof The construction makes $\overline{PQ} \perp \overline{OP}$. PQ is a tangent because the segment is perpendicular to the radius at the point the radius hits the circle.

EXAMPLE 8—

Locate the center of a given circle.

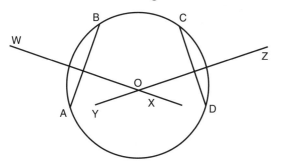

Given: A circle.

Find: the center of the circle.

Construction

1. Draw chords $\overline{AB}$ and $\overline{CD}$. Any nonparallel chords will do.

2. Construct $\overline{DE}$ and $\overline{FG}$, the ⊥ bisectors of $\overline{AB}$ and $\overline{CD}$, respectively.

3. Point O, the intersection of the two perpendicular bisectors, is the circle's center.

Paragraph Proof   The two perpendicular bisectors of the chords (or any chords) go through the center of a circle. Since O is the point where they meet, O must be the circle's center.

NOTE

This theorem can be used in two popular algebraic problems. You would be given the coordinates of A, B, C, and D. Find the slope and midpoint of both $\overline{AB}$ and $\overline{CD}$. Find the equation of the perpendicular bisector of these two chords. Solving two equations in two unknowns locates the center of the circle. Whew! Buuut, we can go even further. Find the length of the radius of the circle, say, $\overline{OA}$. With the radius and center of the circle known, you can write the equation of the circle.

EXAMPLE 9—

Trisect a line segment.

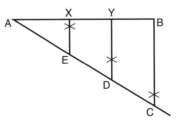

Given: Line segment $\overline{AB}$.

Construct: the trisection of the line segment [three congruent (equal, =) parts].

Construction

1. Draw $\overline{AC}$ at any convenient angle to $\overline{AB}$.

2. On $\overline{AC}$, mark off three lengths that are the same: $\overline{AE} \cong (=) \overline{ED} \cong (=) \overline{DC}$.

3. Draw $\overline{BC}$.

4. Through D and E, construct segments parallel to $\overline{BC}$.

5. $\overline{AX}$, $\overline{XY}$, and $\overline{YB}$ are all the same length.

Paragraph Proof The construction makes $\overline{AE}$, $\overline{ED}$, and $\overline{DC}$ the same length and $\overline{XE}$, $\overline{YD}$, and $\overline{BC}$ parallel. $\overline{AX}$, $\overline{XY}$, and $\overline{YB}$ must be the same since three or more parallel lines cutting off congruent (equal) segments on one transversal must cut off three or more equal (congruent) segments on any transversal.

NOTE 1
You can divide a line into any number of segments by marking off on $\overline{AC}$ any number of equal segments.

NOTE 2
Except for special angles, you can NOT trisect an angle using a ruler and compass only. It is really nice to know that in math, you can prove something can't be done. This means you don't have to waste time trying to do something that is impossible.

Are you OK so far? Let's do the ones that are the most fun.

   Let us construct regular (all sides and angles the same) polygons inscribed in a circle.

EXAMPLE 10

Construct a regular hexagon inscribed in a circle.

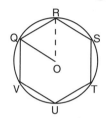

Given: Circle, center O.

Construct: a regular hexagon inscribed in a circle.

Construction

1. Draw radius $\overline{OQ}$.

2. Using A as the center and $\overline{OQ}$ as the radius, draw an arc on the circle, intersecting at R.

3. With the same radius, repeat the procedure at R, S, T, and U, intersecting the circle at S, T, U, and V, respectively.

4. Draw $\overline{QR}$, $\overline{RS}$, $\overline{ST}$, $\overline{TU}$, $\overline{UV}$, and $\overline{VQ}$.

5. QRSTUV is the resulting hexagon.

Paragraph Proof $\overline{OQ} \cong (=) \overline{QR} \cong (=) \overline{OR}$. $\triangle OQR$ is equilateral. $\angle ROQ$ and arc OR are 60°. The circle has been divided into 6 equal arcs. QRSTUV is a regular hexagon because if a circle is divided into any number of congruent (arcs), the chords of these arcs form a regular inscribed polygon.

EXAMPLE 11

Construct an equilateral triangle in a circle.

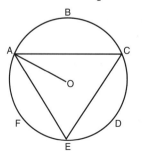

Given: Circle, center O.

Construct: an equilateral triangle inscribed in a circle.

Construction

1. Draw radius $\overline{OA}$.

2. Repeat procedure of example 10, connecting every other arc.

3. $\triangle ACE$ is equilateral.

Paragraph Proof It is the same as example 10. Since arcs AC, CE, and EA are the same, so are the corresponding chords.

EXAMPLE 12

Inscribe a square in a circle.

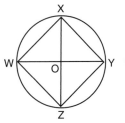

Given: Circle, center O.

Construct: a square inscribed in a circle.

Construction

1. Draw the diameter $\overline{WY}$ for the circle, center at O.

2. At O, construct the perpendicular bisector to $\overline{WX}$, as in example 1.

3. Draw $\overline{WX}$, $\overline{XY}$, $\overline{YZ}$, and $\overline{ZW}$.

4. WXYZ is the required square.

Paragraph Proof Since central angles WOX, XOY, YOZ, and ZOW are right angles, the arcs WX, XY, YZ, and ZW, all 90° are all congruent (equal). The chords must all be congruent (equal). WXYZ is a regular polygon, in this case a square.

EXAMPLE 13—

Construct an octagon inscribed in a circle.

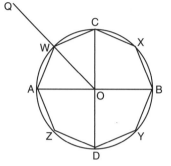

Given: Circle, center O.

Construct: an octagon inscribed in a circle.

Construction

1. Draw diameter $\overline{AB}$.

2. Bisect the diameter, labeling the points C and D where the bisector hits the circle.

3. Bisect ∠AOC, labeling the point that the bisector hits the circle with the letter W.

4. With the compass set at the length of $\overline{AW}$ and using C, B, and D as center, mark each arc, labeling the points X, Y, and Z, where the arcs hit the circle.

5. AWCXBYDZ is the regular octagon.

Paragraph Proof   Since all central angles are the same, so are the arcs and corresponding chords. The figure is a regular octagon.

Let's do one more for fun. We will construct a pentagon, without proof. The reader is encouraged to send in a proof. I will publish the name, city, and state at the first new edition, if my publisher says it is OK.

EXAMPLE 14—

Construct a regular pentagon inscribed in a circle.

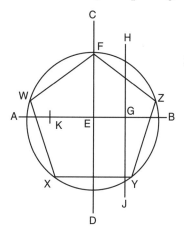

Given: Circle, center E.
Construct: a regular pentagon inscribed in a circle.

Construction

1. Draw diameter $\overline{AB}$ through the center E.

2. Bisect the diameter. The bisector $\overline{CD}$ intersects the diameter at point E, the center of the circle.

3. Bisect radius $\overline{EB}$ at point G, and the bisector hits the circle at point F.

4. With the point of the compass at G and the pencil at F, draw an arc that hits the diameter at point K.

5. With the point of the compass at F and the pencil at K, draw an arc that hits the circle at W.

6. Keeping the radius the same and using W as a center, draw an arc hitting the circle at X.

7. Repeat the procedure using X and Y as a center, labeling the points at which they hit the circle Y and Z.

8. Draw chords $\overline{FW}$, $\overline{WX}$, $\overline{XY}$, $\overline{YZ}$, and $\overline{ZF}$.

9. FWXYZ is a regular pentagon.

In the future, if more constructions are added to this course, this chapter will be extended.

CHAPTER 19

INDIRECT PROOFS, DISPROVING BY COUNTEREXAMPLE; AND TOO MUCH, JUST ENOUGH, NOT ENOUGH

Up to this point we have only done **direct proofs**. This means that we have a given, and in a series of logical steps, we prove the conclusion.

An **indirect proof** is different. We list all the possibilities. Then we prove that all but one is wrong. The remaining choice must be correct.

For example, suppose there are four possibilities: a, b, c, and d. I prove a, b, and d are wrong. Then c must be correct.

Usually there are only two possibilities.

Here are the steps:

1. Draw the picture as before.

2. Write the given and prove as before.

3. Write all the possibilities.

4. In paragraph form, prove all but one of the possibilities wrong. The remaining one must be correct.

Not too bad...

Let's do a set of proofs, some direct and some indirect, that is part of geometry. The facts are needed for the new SAT and several other courses.

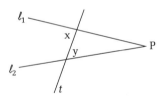

Theorem   If two lines form equal (congruent) alternate interior angles, the lines are parallel.

Given: transversal t, lines ℓ_1 and ℓ_2, (m) $\angle x$ = (m) $\angle y$

Prove: $\ell_1 \parallel \ell_2$

The only possibilities are that either the lines are parallel or the lines are not parallel. Suppose they are not parallel. Then the lines must meet in some point P and we form a triangle. But if this happens (m) $\angle x >$ (m) $\angle y$ since the exterior angle is larger than any remote interior angle. However this is impossible because we assumed that (m) $\angle x$ = (m) $\angle y$. Therefore assuming the lines are not parallel must be wrong. The only other possibility is $\ell_1 \parallel \ell_2$. A really good book has a chapter on proofs like this one.

DEFINITION

Lemma   A theorem (usually smaller than the main theorem) needed to prove a main theorem. If we have two of them, we have a … di-lemma, a dilemma (a joke!!).

DEFINITIONS

Even integer   2n, where n is any integer.

Odd integer   2n + 1, where n is any integer.

Lemma 1   The square of an even integer is even.

Proof $(2n)^2 = 4n^2 = 2(\text{integer})$. Since 2 times an integer is an even integer, we have shown that the square of an even integer is even.

Lemma 2   The square of an odd integer is odd.

Proof $(2n + 1)^2 = 4n^2 + 4n + 1 = 2(2n^2 + 2n) + 1 = 2(\text{integer}) + 1$.

Since the square of an odd integer is 2 times an integer plus 1, the square is odd.

Lemma 3 (indirect)   If the square of an integer is even, then the integer originally was even.

Proof (indirect)   If the square of an integer is even, then originally the integer was even or odd.

Suppose the integer was odd. By lemma 2, the square must be odd. This is false since the square is even. Therefore, the original integer couldn't be odd. It must be even.

   (Please notice I've added extra words for clarity. I hope I haven't added too many.)

Theorem $\sqrt{2}$ is irrational.

Proof (indirect)   Either the $\sqrt{2}$ is rational or irrational. Suppose it is rational. Then

$$\sqrt{2} = \frac{a}{b}$$

where a and b are integers, $b \neq 0$, and a/b is reduced to lowest terms.

   Then, squaring both sides,

$$2 = \frac{a^2}{b^2} \quad \text{or} \quad a^2 = 2b^2$$

This means a^2 is even. By lemma 3, this means a is even. Soooo, a = 2x where x is an integer:

$$2b^2 = a^2 = (2x)^2 = 4x^2 \qquad b^2 = 2x^2$$

So b^2 is even. Again by lemma 3, b is even.

   So b = 2y, where y is an integer. Buuut a/b = 2x/2y. We can cancel the 2s. However, that is impossible because we assumed a/b was in lowest terms (we already did all the canceling). Assuming that the $\sqrt{2}$ is rational must be false; therefore, $\sqrt{2}$ is irrational.

DISPROVING BY COUNTEREXAMPLE

This is perhaps the easiest part of the course.

To disprove a statement, all you do is find one example (counterexample) to show the statement is wrong.

EXAMPLE 1

No right triangle is isosceles.

Counterexample   A triangle whose angles are 45°, 45°, 90°. (In this case, it is the only counterexample.)

EXAMPLE 2

All rectangles are squares.

Counterexample   A rectangle whose sides are 5, 7, 5, and 7. (There are many counterexamples.)

Now if the question is presented as it is given in examples 1 and 2, it is not too hard. However, you might get problems like these:

1. If there are not enough hypotheses, give a counterexample.

2. If there is just enough information, prove the theorem.

3. If there is too much information, eliminate whatever you can and prove the theorem.

EXAMPLE 3

If two sides of a triangle are equal (congruent), the angles opposite those sides are equal (congruent).

This is a statement on isosceles triangles that is true.

Outline of Proof Draw an isosceles triangle. Bisect the vertex angle. Triangles congruent by SAS.

Base angles are equal (congruent) by corresponding parts.

EXAMPLE 4—

If two angles of a triangle are equal (congruent) to two angles of another triangle, the triangles are congruent.

Counterexample There are lots. Two angles means that the triangles are similar. One triangle bigger than the other but the same shape is a counterexample.

EXAMPLE 5—

Given: △ABC and △XYZ. And suppose that ∠A = (≅) ∠X, ∠B = (≅) ∠Y, $\overline{BC}$ = (≅) $\overline{YZ}$, $\overline{AB}$ = (≅) $\overline{XY}$. *Prove:* that the two triangles are congruent.

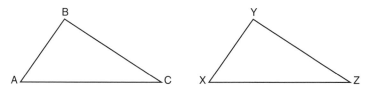

Explanation This is toooo much information. Draw the triangles.

1. Eliminating $\overline{BC}$ ≅ $\overline{YZ}$, the triangles are congruent ASA.

2. Eliminating $\overline{AB}$ ≅ $\overline{XY}$, the triangles are congruent by AAS if allowed, or you can say the third angles are the same with the triangles congruent by ASA.

3. Eliminating ∠A ≅ ∠X and the triangles are congruent by SAS.

4. You cannot eliminate the other angle because triangles are NOT congruent by SSA.

NOTE
You could be given too much information that contradicts itself. So you would have to show the statement is not true by counterexample.

Questions like these require you to know everything verrrry well!!!!

CHAPTER 20

MISCELLANEOUS: LOCUS; PARALLEL LINES; AND LARGER AND SMALLER SIDES AND ANGLES

The following topics do not fit into a particular section easily. So let's do them here.

LOCUS

My heart feels bad because these problems are rarely done anymore. But let's do a few of these here.

DEFINITION

Locus A point or the set of all points and only those points satisfying a condition or conditions. Some examples will explain this. You should draw the picture if possible.

EXAMPLE 1—

The set of all points at a distance r from a given point P.

One dimension: Draw the line through P (that is the one dimension): The locus is two points, distance away from P.

Two dimensions (x-y plane) locus: Circle, radius r, center P.

Three dimensions (space) locus: Sphere, radius r, center P.

EXAMPLE 2—

The set of all points equidistant from two points, A and B.

One dimension: Draw the line segment $\overline{AB}$. The midpoint.

Two dimensions: The perpendicular bisector of $\overline{AB}$.

Three dimensions: Draw $\overline{AB}$. Draw the perpendicular bisector of $\overline{AB}$. Draw the plane through $\overline{AB}$ and its perpendicular bisector (remember, there is only one). *Locus:* A plane through the perpendicular bisector and perpendicular to the plane you just drew. (3-D can be hard to visualize.)

EXAMPLE 3—

The set of all points at a distance r from line a parallel to line a.

One dimension: Impossible since the line is the one dimension.

Two dimensions: Two lines parallel to line a at a distance r from the line.

Three dimensions: A cylinder with line a through its center, radius of the circular base r.

EXAMPLE 4—

The set of all points equidistant from the sides of an angle.

One dimension: Impossible since angles are two dimensional.

Two dimensions: The bisector of the angle.

Three dimensions: The plane perpendicular to the plane the angle is in (an angle is alllways in one plane) through the bisector of the angle.

EXAMPLE 5

(This is a toughie.) The set of all points at a distance a from a line segment. This is a combination of examples 1 and 3.

One dimension: Two points, P and Q, one at a distance a from the left end and one at the right end.

Two dimensions: Two line segments parallel to the line segment at a distance a plus a semicircle of radius a at each end. Oh, let's draw the picture!

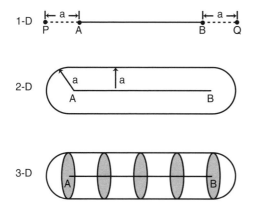

Three dimensions: A cylinder the length of the line segment, radius a plus two hemispheres (half-spheres) are the ends of the line segment, radius a.

NOTE
The locus of points in a plane equidistant from a point and a line is a parabola.
We discussed this in the section "Parabola" in Chapter 16.

Let's go on.

PARALLEL LINES

Theorem I

If three or more parallel lines are cut by two transversals, the lines make proportional segments on each transversal.

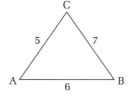

EXAMPLE 6—

The proportion is x/5 = 7/8.

$$8x = 35 \qquad x = 35/8.$$

They can be made a little harder, but not much.

I knew I forgot something!

LARGER AND SMALLER SIDES AND ANGLES

Theorem 2

In a triangle, the largest side lies opposite the largest angle and the largest angle lies opposite the largest side.

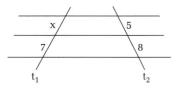

EXAMPLE 7—

The largest angle is A because it is opposite the longest side, 7. Next is angle C. The smallest is angle B.

EXAMPLE 8—

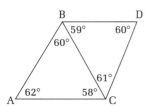

Which is the largest side? In △ABC, the largest side is BC since it is opposite the largest angle, 62 degrees. In △BCD, the largest side is $\overline{BD}$ since it lies opposite the 61 degree angle in the second triangle.

Buuuuut $\overline{BC}$ and $\overline{BD}$ are in the same triangle!!! Since $\overline{BD}$ is the largest side in this triangle, $\overline{BD}$ is the largest side.

Actually this section is slightly out of place since in trig the law of sines tells you exactly the relationship between angles and sides. You will learn more about the law of sines if you study trigonometry.

CHAPTER 21

TWENTY-FIRST-CENTURY SAT SPIN ON GEOMETRY

In 2004, the College Board instituted its second major set of changes on the SAT. (The first was 1993.) They started with the March 2005 test. Fortunately most of those topics are already in this book. The algebraic chapter are not appropriate for this book. They can be found in the third edition of *Pre Calc for the Clueless* and *SAT® Math for the Clueless,* Second Edition. We will cover one major topic—sets—and one minor topic—tangent to a circle—in this chapter. Then we'll give SAT-like problems.

SETS

We need some basic terms.

A **set** is a collection of things called **elements**.

A capital letter, such as S, indicates a set. A lower-case letter indicates an element.

When you write a set, you use braces: { }.

For example, {2, 5, 6} is a set that has three elements: 2, 5, and 6.

$6 \in \{2, 5, 6\}$ means that 6 is an element in the set.

NOTE

( ) are parentheses.

[ ] are brackets.

$\in$ is the Greek letter epsilon.

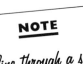

7 ∉ {2, 5, 6} means 7 is **not** an element in the set.

∅ or { } means the null set or empty set.

For example, the set of all 25-foot human beings is the null set.

WARNING!

{0} is a set with one element in it, the number zero. It is not, not, not the null set!

A ∪ B, read "A union B," is the set of all elements in A or in B or in both.

A ∩ B, read "A intersect B" or "A intersection B," is the set of all elements in both sets.

EXAMPLE 1—

Let A = {1, 2, 4, 7}, B = {2, 5, 7, 8}, and C = {3, 5, 6}.

Then A ∪ B = {1, 2, 4, 5, 7, 8}. A ∩ B = {2, 7}.
A ∩ C = ∅. Such sets are called **disjoint**.

EXAMPLE 2—

A = {3, 5}   B = {5, 3}   C = {3, 3, 5, 3, 5, 5, 3, 5}

A = B since order does not matter in a set.

A = C since repeated elements count only once. Notice set C has only two elements!!! We say A is a subset of B, written A ⊆ B, if every element in A is also in B.

EXAMPLE 3—

{4, 7} ⊆ {7, 1, 4, 100}

If there are n elements in a set, then there are 2^n subsets.

EXAMPLE 4—

List all the subsets of {a, b, c}.

The set has three elements. There are 2^3, or 8, subsets. They are ∅, {a}, {b}, {c}, {a, b}, {a, c}, {b, c} and {a, b, c}.

Let's try some questions.

EXAMPLE 5—

A = {1, 2, 3, 4}   B = {3, 4, 5}   A ∪ B = what?

a. {3, 4}

b. {1, 2, 3, 4, 5}

c. {4, 5, 6, 7, 8, 9}

d. ∅

e. None of these

EXAMPLE 6—

A = {5, 6, 7, 8, 9}   B = {2, 3, 4, 5, 6}.   A ∩ B =

a. {5, 6}

b. {2, 3, 4, 5, 6, 7, 8, 9}

c. {2, 3, 4, 5, 5, 6, 6, 7, 8, 9}

d. {10, 18, 28, 40, 54}

e. ∅

EXAMPLE 7—

The union of the odd integers and the even integers is

A. All real numbers

B. All counting numbers

C. All integers

D. All rational numbers

E. ∅

EXAMPLE 8—

The intersection of all nonnegative integers and all nonpositive integers is

A. ∅

B. { }

C. {0}

D. All integers

E. All real numbers

EXAMPLE 9—

The intersection of the x axis and the y axis is

A. All points (x, 0) or (0, y)

B. The x-y plane

C. All ordered pairs

D. The origin

E. None of these

EXAMPLE 10—

Given {a, b, c, d}, how many subsets are there not counting the null set and the set itself?

A. 18

B. 17

C. 16

D. 15

E. 14

Answers

EXAMPLE 5 ANSWER—

b. *Union* means put all elements that have appeared once in one set.

EXAMPLE 6 ANSWER—

a. *Intersection* means common to both sets.

EXAMPLE 7 ANSWER—

C

EXAMPLE 8 ANSWER—

C. Nonnegative integers mean 0, 1, 2, 3, 4,...

Nonpositive integers mean 0, −1, −2, −3,...

They both have zero in common.

EXAMPLE 9 ANSWER—

D.   They meet at the origin (0, 0).

EXAMPLE 10 ANSWER—

EEEEE. Four elements. 2^4, or 16, subsets. $16 - 2 = 14$.

TANGENT TO A CIRCLE

As we learned before, a tangent to a circle is a line that hits the circle in one and only one point. A fact only briefly mentioned in Chapter 11, theorem 11, is that the radius drawn to the circle is perpendicular to the tangent at the point of tangency.

There are a lot of nice problems based on this. Let's do two, one easy and one not so easy.

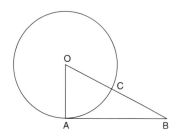

Given: Circle, center at O. A is the point of tangency. $\overline{AB}$ is tangent to the circle. $\overline{OB}$ intersects the circle at C.

EXAMPLE 11

The diameter of the circle is 12, and $\overline{AB}$ is 8. Find $\overline{OB}$.

EXAMPLE 12

$\overline{AB} = 4$ and $\overline{BC} = 2$. Find the diameter of the circle.

Answers

EXAMPLE 11 ANSWER

The diameter is 12; so the radius $\overline{OA}$ is 6. $\overline{AB}$ is 8. $\overline{OB}$, the hypotenuse, is 10 because we have a 6–8–10 right triangle.

EXAMPLE 12 ANSWER

$\overline{OA}$ and $\overline{OC}$ are both radii. Let them each equal x, because all radii are congruent. By the Pythagorean theorem:

$$x^2 + 4^2 = (x + 2)^2$$

$$x^2 + 16 = x^2 + 4x + 4.$$

The x^2 terms cancel:

$$4x + 4 = 16 \qquad 4x = 12; x = 3.$$

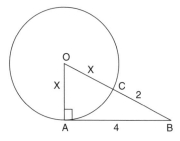

The radius x is 3. Sooo, the diameter is 6.

SAMPLE SAT QUESTIONS

The SAT asks geometry questions in different ways than your teachers. In this section we'll do a few. Most of the SAT is multiple choice. These will be "open ended." You have to supply the answer.

EXAMPLE 13—

Let A = {1, 2, 3, 4, 5, 6}; B = {1, 3, 4, 5, 6, 7}; C = {1, 6, 7, 8, 9}. (A ∩ B) ∪ (A ∩ C) = what?

EXAMPLE 13 ANSWER—

A ∩ B = {1, 3, 4, 5}, and A ∩ C = {1, 6}. The union of these two sets is {1, 3, 4, 5, 6}.

EXAMPLE 14—

Let the point (1, 2) be on f(x):

A. On the curve f(x + 2), (1, 2) transforms into what point?

B. On the curve f(x − 3), (1, 2) transforms into what point?

C. On the curve 5f(x), (1, 2) transforms into what point?

D. On the curve f(x) + 5, (1, 2) transforms into what point?

E. On the curve −f(x), (1, 2) transforms into what point?

EXAMPLE 14 ANSWERS—

A. Two to the left: (−1, 2)

B. Three to the right: (4, 2)

C. Five times as high: (1, 10)

D. Five up (1, 7)

E. Upside down (y becomes −y), (1, −2).

EXAMPLE 15—

Find x.

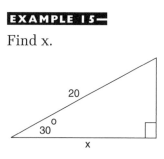

EXAMPLE 16—

Diameter $\overline{AB}$ = 20. Find the area of the shaded portions.

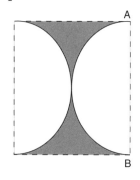

EXAMPLE 17—

Given a circle center O, $\overline{OA} = \overline{AB} = \overline{OB} = \overline{AE} = \overline{BF}$. $\overline{CD} = 12$.

Find the area of the shaded portion.

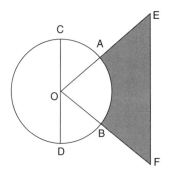

Answers

EXAMPLE 15 ANSWER—

We have a 30°, 60°, 90° right triangle. Side 20 is the hypotenuse. The short leg, opposite the 30° angle, is half the hypotenuse, 10. The long leg, opposite the 60° angle, is the short leg times $\sqrt{3}$ or $10\sqrt{3}$, approximately 17.32. You can use right angle trig:

$$cos\ 60° = \frac{\text{adjacent}}{\text{hypotenuse}} = \frac{x}{20}$$

So x = 20 cos 30° ≈ 17.32 by calculator. I think I prefer the first way, but you have the last word.

EXAMPLE 16 ANSWER—

You must see that the shaded portion is a square minus a circle (two half circles). The formula is $s^2 - \pi r^2$ because d = 20, the r = 10, and s = 20:

$$A = 20^2 - \pi (10)^2 = 400 - 100\pi$$

Do not factor the answer or change to a decimal unless the multiple-choice answer does. The SAT never used to ask questions like this on the SAT because everyone used to know how to do them. They were found in all geometry books. In fact I found the two books I used as a student. The books are filled with these problems. Most of you have seen very few.

Example 17 is another example of the same.

EXAMPLE 17 ANSWER—

First you must notice that △BOA is an equilateral triangle. Then so is △OEF. This also means that all angles are 60°.

Next we notice that $\overline{COD}$ is a diameter of the circle. $\overline{COD} = 12$, so the radius $= 6$. We notice the shaded portion is an equilateral triangle, side 12, minus a sector of the circle. This sector is 60°/360° or 1/6 the area of the circle.

The shaded area $= \dfrac{s^2\sqrt{3}}{4} - \dfrac{1}{6}\pi\, r^2 = \dfrac{12^2\,\sqrt{3}}{4} - \dfrac{1}{6}\pi(6)^2$

$$= 36\sqrt{3} - 6\pi.$$

In order to do this problem quickly, you must have memorized the formulas for the area of a circle and equilateral triangle.

If you want more problems, go to my *SAT® Math for the Clueless*, Second Edition. For now, let's try some always-sometimes-never questions because most geometry teachers give these kinds of questions.

Let's do some always-sometimes-never questions.

CHAPTER 22

ALWAYS-SOMETIMES-NEVER QUESTIONS

To help you find out how well you know geometry and because teachers very often ask these questions, we will do always-sometimes-never questions. You must fill in one of these choices in the blank space in the questions.

Always means always true.

Sometimes means true at least one time and false at least one time.

Never means always false.

Let's go!!!!!

INTRODUCTORY STUFF

1. In a plane, a series of connected line segments is _____ a polygon.

2. A bisector of a line segment _____ passes through the midpoint of the line segment.

3. When two lines intersect, all the vertical angles are _____ equal (congruent).

4. The complement of an acute angle is _____ acute.

5. The supplement of an acute angle is _____ acute.

6. If two angles are supplementary, they are _____ equal (congruent).

7. In an isosceles triangle, base angles are _____ obtuse.

8. If $\overline{WX}$ is the perpendicular bisector of $\overline{YZ}$, then $\overline{YZ}$ _____ is the perpendicular bisector of $\overline{WX}$.

9. Two angles with a common vertex are _____ adjacent angles.

10. The bisectors of two adjacent supplementary angles are _____ perpendicular.

11. If equals are multiplied by equals, their products are _____ equal.

12. If two lines meet, their vertical angles are _____ supplementary.

13. A theorem is a statement that is _____ assumed to be true.

14. Two complementary angles are _____ equal (congruent).

15. Two lines that don't intersect are _____ parallel.

16. Three points _____ determine exactly one plane.

17. Two intersecting lines _____ determine exactly one plane.

18. Two parallel lines _____ determine exactly one plane.

19. Two skew lines _____ determine exactly one plane.

20. Two rays _____ intersect.

21. Two angles are _____ adjacent if they have a common side.

22. If $\overline{WX}$ is perpendicular to $\overline{YZ}$, then $\overline{YZ}$ is _____ perpendicular to $\overline{WX}$.

23. If $\overline{WX}$ bisects $\overline{YZ}$, $\overline{YZ}$ _____ bisects $\overline{WX}$.

24. If a theorem is true, its converse is _____ false.

25. If a theorem is true, its contrapositive is _____ false.

26. If we have an "if and only if" theorem, the statement and the converse are _____ true.

27. If the converse is false, the inverse is _____ true.

28. The converse of the inverse is _____ the contrapositive.

29. A counterexample can _____ prove a statement false.

CONGRUENT AND OTHER TRIANGLES

30. Two triangles are _____ congruent if three angles of one are equal (congruent) to three angles of the other.

31. Two triangles are _____ congruent if two sides and an angle of one are =($\cong$) to two sides and an angle of the other.

32. Two triangles are _____ congruent if two angles and a side of one are =(≅) to two angles and a side of the other.

33. Two triangles are _____ congruent if three sides of one are equal (congruent) to three sides of the other.

34. Two right triangles are _____ congruent if the hypotenuse and one leg of one is equal (congruent) to the same in the other.

35. Two right triangles are _____ congruent if one leg and one acute angle are =(≅) to the same in the other.

36. Two isosceles triangles are _____ congruent if the vertex and one base angle of one are =(≅) to the same in the other.

37. Two isosceles triangles are _____ congruent if one leg and the vertex angle of one are =(≅) to the same in the other.

38. In an isosceles triangle two medians are _____ equal (congruent).

39. The altitude to the base of an isosceles triangle _____ bisects the vertex angle.

40. The altitude to the base of an isosceles triangle _____ bisects the base.

41. An equilateral triangle is _____ equiangular.

42. Two triangles are _____ similar if two angles of one △ are = (≅) to two angles of the other △.

43. A median of a triangle _____ hits the midpoint of one of the sides of a triangle.

44. The bisector of the angle of an isosceles triangle _____ divides the triangle into two congruent triangles.

QUADS, PARALLELS, PERPS, AND ANGLE SUMS

45. Two lines cut by a transversal _____ have alternate interior angles equal (congruent).

46. Two parallel lines cut by a transversal _____ have interior angles on the same side of the transversal =($\cong$).

47. The bisector of interior angles on the same side of the transversal of parallel lines are _____ =($\cong$).

48. The bisector of interior angles on the same side of the transversal of parallel lines are _____ $\perp$.

49. Alternate exterior angles of two parallel lines are _____ supplementary.

50. If a line is parallel to one of two parallel lines, it is _____ parallel to the other.

51. If a line is parallel to one of two perpendicular lines, it is _____ perpendicular to the other.

52. The measure of an exterior angle of a polygon is _____ 80°.

53. If the diagonals of a quadrilateral are equal (congruent), the figure is _____ a rhombus.

54. If the diagonals are equal (congruent), the quadrilateral is _____ a rectangle.

55. If the diagonals of a quadrilateral bisect the angles of the figure, the figure _____ is a rhombus.

56. If the diagonals of a quadrilateral form four congruent triangles, it is _____ a rhombus.

57. A quadrilateral with four right angles is _____ a square.

58. The bisector of the opposite angles of a parallelogram _____ coincide.

59. The sum of the measures of the interior angles of a quadrilateral is _____ the same as the sum of the measures of the exterior angles.

60. The diagonals of a trapezoid are _____ equal (congruent).

61. If two sides of a quadrilateral are equal (congruent) and two are parallel, the figure is _____ a parallelogram.

CIRCLES

62. A chord is _____ a diameter.

63. A chord is _____ a radius.

64. A diameter is _____ a chord.

65. An angle formed by two tangents is _____ equal in degrees to half the measure of the intercepted arc.

66. If two arcs of a circle are equal (congruent), their chords are _____ equal (congruent).

67. The perpendicular bisector of a chord _____ passes through the center of the circle.

68. An angle inscribed in a semicircle is _____ a right triangle.

69. A parallelogram inscribed in a circle is _____ two acute angles.

70. If two angles intercept the same arc, they are _____ equal (congruent).

71. If a chord in one circle is equal (congruent) to a chord in another circle, the minor arcs they intercept are _____ equal (congruent).

72. If two circles are concentric (having the same center), any chords of the larger circle that are also tangent to the smaller circle are _____ equal (congruent).

73. Any angle inscribed in an arc less than a semicircle is _____ obtuse.

74. The part of a secant that hits the circle is _____ a chord.

75. The part of a tangent that hits the circle is _____ a chord.

76. A tangent to a circle is _____ perpendicular to the radius at the point of tangency.

77. Two tangents from the same external point are ____ = ($\cong$).

SIMILAR TRIANGLES, SEGMENTS, LOCUS, AND A LITTLE ON TRIG

78. Congruent polygons are _____ similar.

79. A median of a triangle _____ divides the triangle into two similar triangles.

80. The sine of an angle _____ is the cosine of its complement.

81. Similar triangles are _____ congruent triangles.

82. Two isosceles triangles are _____ similar if their vertex angles are equal (congruent).

83. Two right triangles are _____ similar if one acute angle of one is equal (congruent) to the other.

84. As an angle increases, the tangent of that angle _____ increases.

85. As an angle increases, the cosine of that angle _____ increases.

86. A rhombus whose side is 10 inches _____ has a diagonal of 20 inches.

87. If the ratio of two sides of two similar polygons is a/b, then the ratio of their perimeters is _____ a/b.

88. If two polygons have corresponding angles equal (congruent), they are _____ similar.

89. Two polygons are _____ similar if all corresponding sides are in proportion.

90. If two sides of a triangle have lengths 4 and 7, the remaining side is _____ 9.

91. If two sides of a triangle have lengths 4 and 7, the remaining side is _____ 2.

92. The locus of all points equidistant from two points is _____ the perpendicular bisector of the line segment joining these two points.

93. The locus of all points in a plane equidistant from a point is _____ a circle.

94. The locus of all points in a plane equidistant from the sides of a triangle is _____ a point.

95. The locus of all points in space equidistant from the sides of a triangle is _____ a line.

96. Two isosceles triangles are _____ similar if their legs are in proportion.

97. Two equilateral triangles are _____ congruent.

AREAS AND VOLUMES

98. If the ratio of the area of two similar figures is 4/9, the ratio of their volumes is _____ 8/27.

99. If two polygons are congruent, they are _____ equal in area.

100. Triangles with the same base and same altitude to that base are _____ congruent.

101. If the ratio of two similar triangles is 1/7, the larger base is _____ 7 times the smaller base.

102. The area of a rhombus is _____ one-half the product of its diagonals.

103. If the sides of two rhombuses are equal (congruent), the areas are _____ equal.

104. If the volumes of two cylinders are equal, the cylinders are _____ congruent.

105. If the radii of two spheres are the same, their volumes are _____ equal.

106. If two cylinders have the radius of the base the same and the height the same, they are _____ congruent.

107. The ratio of their area of two unequal circles is _____ the same as the ratio of their radius.

108. If the bases of a triangle are the same, the ratio of their areas are _____ the ratio of the altitudes drawn to that base.

109. A cylinder and a cone have the same height and the same radius of the base. The volume of the cylinder is _____ three times the volume of the cone.

110. The volume of a cube is numerically _____ the same as its surface area.

This is just a sampling. Let's take a look at the answers.

CHAPTER 23

ANSWERS: YOU SHOULD ALWAYS DRAW THE PICTURE IF YOU ARE HAVING PROBLEMS

1. Sometimes. Only when it is closed.

2. Always.

3. Sometimes. When they are perpendicular.

4. Always.

5. Never. Always obtuse.

6. Sometimes. Again when they are right angles.

7. Never. The sum of the angles must be 180°; two obtuse angles would total more than 180°.

8. Sometimes. Allllways perpendicular buut sometimes bisect.

9. Sometimes. When there is a common side between.

10. Always.

11. Always. An axiom.

12. Sometimes. When they are perpendicular.

13. Never. Theorems always must be proven, even if we don't do them; postulates and axioms are assumed to be true.

14. Sometimes. When they are each 45°.

15. Sometimes. If it is in a plane, it would be always.

16. Sometimes. If they are colinear, it would be an infinite number.

17. Always.

18. Always.

19. Never.

20. Sometimes.

21. Sometimes. If the question said the side is between, the answer would be always.

22. Always.

23. Sometimes

24. Sometimes.

25. Never.

26. Always.

27. Never.

28. Always.

29. Always.

30. Sometimes. Always similar but can be bigger or smaller.

31. Sometimes. Only if the angle is between the sides.

32. Always. Two angles mean three angles and we allllways have ASA.

33. Always.

34. Always. HL.

35. Always. ASA.

36. Sometimes. The same angles mean similar, but they are not necessarily congruents, because at least one pair of congruent sides is necessary.

37. Always. SAS.

38. Always. Medians from each base angle.

39. Always.

40. Always.

41. Always.

42. Always.

43. Always.

44. Sometimes. It would be always if the problem said the vertex angle.

45. Sometimes. They must be parallel!!!!!! These questions can be verrrrry tricky.

46. Sometimes. When the transversal is perpendicular to them.

47. Same answer as for question 46: Sometimes. When the transversal is perpendicular to them.

48. Always!! Draw the picture.

49. Sometimes. See 46 and 47.

50. Always.

51. Sometimes. Always if they are all in one plane.

52. Never. 360 is not divisible by 80.

53. Sometimes. If it is also a square.

54. Sometimes. It might be an arrowhead or an isosceles trapezoid.

55. Always.

56. Always.

57. Sometimes. It might only be a rectangle.

58. Sometimes. When it is also a rhombus or a square.

59. Always.

60. Sometimes. When it is also isosceles.

61. Sometimes. It might be an isosceles trapezoid.

62. Sometimes.

63. Never.

64. Always. The biggest chord.

65. Sometimes; really tough question.

66. Always.

67. Always. Draw the picture.

68. Never. An angle inscribed in a semicircle is a right angle.

69. Never. It must always be at least a rectangle. Again, draw it.

70. Sometimes.

71. Sometimes.

72. Always.

73. Never. It is acute since half of less than 180° is less than 90°.

74. Always.

75. Never.

76. Always.

77. Always.

78. Always.

79. Sometimes.

80. Always.

81. Sometimes.

82. Always.

83. Always.

84. Always.

85. Never.

86. Never.

87. Sometimes. The sides must be corresponding for the answer to be always.

88. Sometimes. The statement would not be true for a square since a square is not similar to a square. The statement is always true only in the case of a triangle.

89. Sometimes. It is always true only in the case of a triangle.

90. Sometimes.

91. Never. If two sides are 4 and 7, the remaining side must be bigger than 3 ($7 - 4$) and less than 11 ($7 + 4$). If the problem said "integers," the answer could be 4, 5, 6, 7, 8, 9, or 10.

92. Always. It is a point in one dimension, a line in 2-D, and a plane in 3-D. They are all perpendicular bisectors (although a point is stretching it).

93. Sometimes. In a plane, it would be always.

94. Always.

95. Always.

96. Sometimes.

97. Sometimes.

98. Always. Ratio of similarity is the square root, 2/3; volumes would be 2/3 cubed or 8/27.

99. Always. Congruent means the same in every way!!!!!

100. Sometimes. Right triangles and not right triangles might have the same area.

101. Always.

102. Always.

103. Sometimes. The angles at their bases could be different; so are their heights and diagonals.

104. Sometimes.

105. Always.

106. Sometimes!!!!! Cylinder can be at an angle!!!

107. Never. Ratio of areas is the square of the ratio of radii (the ratio of similarity).

108. Always.

109. Always.

110. Sometimes.

As you can tell, some of these are easy and some are not so easy, but all of them make you think and know your geometry. Because they make you think and because they are short answers are the reasons teachers ask these questions. Perhaps your book has a few more for you to practice.

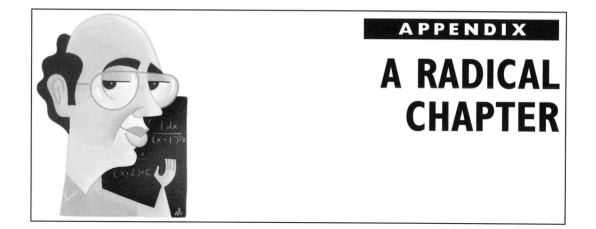

A RADICAL CHAPTER

When I started teaching, I thought that this topic caused students many problems. Maybe it was me, and maybe I was wrong, but now my students learn this topic super well. I'm sure you will too.

DEFINITION OF RADICALS

$\sqrt{a}$, "square root of a": the number b, such that (b)(b) = a.

$\sqrt{9} = 3$, because (3)(3) = 9. $\sqrt{25} = 5$ because (5)(5) = 25. 9 and 25 are called *perfect squares* because they have exact square roots.

$\sqrt[3]{a}$, "cube root of a": the number c, such that (c)(c)(c) = a. $\sqrt[3]{64} = 4$ because (4)(4)(4) = 64.

$\sqrt[4]{a}$, "the fourth root of a": the number h, such that (h)(h)(h)(h) = a. $\sqrt[4]{81} = 3$ because (3)(3)(3)(3) = 81.

$\sqrt[n]{a}$, "the nth root of a": the number b such that
(b)(b)(b) . . . (b) = a
(n factors).

EXAMPLE 1—

$\sqrt{16} = 4$; $-\sqrt{16} = -4$ (minus sign is on the outside of square root).

$\pm\sqrt{16} = \pm4$ ($\pm\sqrt{16}$ means two numbers, $+\sqrt{16}$ and $-\sqrt{16}$).

$\sqrt{-16}$ is not a real number because a positive times a positive is positive and a negative times a negative is positive. No real number squared gives a negative. Also, $\sqrt{0} = 0$. Buuuut . . .

EXAMPLE 2—

$\sqrt[3]{1000} = 10$; $\sqrt[3]{-125} = -5$; $\sqrt[3]{0} = 0$

$-\sqrt[3]{-27} = -(-3) = 3$; $\pm\sqrt[3]{216} = \pm6$; $\pm\sqrt[3]{-8} = \pm2$

(The chart is on the next page.)

As you do these problems, you should keep this list in front of you. Don't go home and memorize. But as you do the problems, you will be surprised how many you learn from repetition.

EXAMPLE 3—

Solve for x:

$x^2 - 16 = 0$

$x^2 = 16$

Two roots.

$x = \pm\sqrt{16} = \pm4$

Powers

	Square	Cube	4th	5th	6th	7th	8th	9th	10th	11th	12th
2	4	8	16	32	64	128	256	512	1024	2048	4096
3	9	27	81	243	729						
4	16	64	256	1024							
5	25	125	625	3125							
6	36	216									
7	49	343									
8	64	512									
9	81	729									
10	100	1000									
11	121	1331									
12	144	1728									
13	169										
14	196										
15	225										
16	256										
17	289										
18	324										
19	361										
20	400										
21	441										
22	484										
23	529										
24	576										
25	625										
26	676										
27	729										
28	784										
29	841										
30	900										
31	961										
32	1024										

The chart is read backward and forward. 5 cubed is
125, buuuut also the cube *root* of 125 is 5.

NOTES

You will find that most of these numbers repeat over and over; some now and some later on.

Try not to use calculators. If you put these numbers in your brain, many things go easier for you.

Some numbers are not perfect squares. 2 is not a perfect square. There is no integer or rational number that gives you 2 when you square it. $\sqrt{2}$ is *irrational*.

An *irrational number* is any number that cannot be written as a terminating or repeating decimal (any number that is not rational).

π, $\sqrt{7}$, and $\sqrt[3]{5}$ are all examples of irrational numbers.

Remember, writing 3.14 or 22/7 are only approximations for π.

A *real* number is any decimal number.

The real numbers are the rationals plus the irrationals.

Now that we know what square and other roots are, we would like to simplify them; add, subtract, multiply, and divide them; and apply them.

Ninety-five percent of the time, the shortest way is the simplest. I was always taught to use the K.I.S.S. method of teaching—keep it simple, stupid. However, the radical unit is one of the few times we don't use the shortest method. The method is short, just not the shortest.

SIMPLIFYING RADICALS

EXAMPLE 1—

Simplify

$\sqrt{18}$

$= \sqrt{2 \boxed{(3)(3)}}$

$= 3\sqrt{2}$

We write 18 as the product of primes: 18 = 2(3)(3).

Two 3s circled on the inside become one 3 on the outside. Why? Because $\sqrt{(3)(3)}$ = $(\sqrt{9})$ = 3.

EXAMPLE 2—

Simplify

$\sqrt{72}$

$= \sqrt{\boxed{(2)(2)}(2)\boxed{(3)(3)}} = 2 \times 3\sqrt{2} = 6\sqrt{2}$

Break 72 into primes. No matter how you do it, you will get the same ones.

EXAMPLE 3—

Simplify

$10\sqrt{125}$

$= 10\sqrt{\boxed{(5)(5)}(5)} = 10(5)\sqrt{5} = 50\sqrt{5}$

EXAMPLE 4—

Simplify

$\sqrt{27b^7} = \sqrt{(3)\boxed{(3)\,(3)}\boxed{(b)(b)}\boxed{(b)(b)}(b)(b)(b)} = 3b^3\sqrt{3b}$

EXAMPLE 5—

Simplify

$\sqrt{a^{11}} = \sqrt{\boxed{aa}\boxed{aa}\boxed{aa}\boxed{aa}\boxed{aa}a} = a^5\sqrt{a}$

Short way!!!! Nice! $\sqrt{}$ means $\sqrt[2]{}$. Divide 2 into 11. We get 5 groups of 2. So a^5 goes on the outside. The remainder is 1. a goes on the inside. That's all!

EXAMPLE 6—

Simplify

$$4y^3\sqrt{25y^3} = (4y^3)(5y)\sqrt{y} = 20y^4\sqrt{y}$$

EXAMPLE 7—

Simplify

$$\sqrt[3]{72a^{14}b^9}$$

With cube roots, we circle three of the same factors. (Fourth roots we circle 4 of the same.) 3 into 14 is 4 with a remainder of 2. 3 into 9 is 3 with no remainder. Soooooo,

$$\sqrt[3]{72a^{14}b^9} = 2a^4b^3\sqrt[3]{(3)(3)a^2} = 2a^4b^3\sqrt[3]{9a^2}$$

Since

$$\sqrt[3]{72} = \sqrt[3]{\boxed{(2)\,(2)\,(2)}(3)(3)} = 2\sqrt[3]{(3)(3)}$$

Not too bad!!

ADDING AND SUBTRACTING RADICALS

Adding and subtracting radicals is the same as adding and subtracting like terms. With like radicals, you add or subtract the coefficients. Unlike radicals cannot be combined.

EXAMPLE 1—

Simplify

$$4\sqrt{2} - 5\sqrt{7} - 7\sqrt{2} - 10\sqrt{7}$$

The answer issss

$$-3\sqrt{2} - 15\sqrt{7}$$

EXAMPLE 2—

Simplify

$3\sqrt{27} + 5\sqrt{8} + 6\sqrt{12}$

$= 3\sqrt{(3)(3)(3)} + 5\sqrt{(2)(2)(2)} + 6\sqrt{(2)(2)(3)}$

$= (3)(3)\sqrt{3} + 5(2)\sqrt{2} + (6)(2)\sqrt{3} = 9\sqrt{3} + 10\sqrt{2} + 12\sqrt{3}$

$= 21\sqrt{3} + 10\sqrt{2}$

We must simplify each radical first.

EXAMPLE 3—

Simplify

$\sqrt[3]{128y^8} + 10y\sqrt[3]{54y^5}$

$= \sqrt[3]{(2)(2)(2)(2)(2)(2)(2)(y)(y)(y)(y)(y)(y)(y)(y)}$

$+ 10y\sqrt[3]{(2)(3)(3)(3)(y)(y)(y)(y)(y)}$

$= 4y^2\sqrt[3]{2y^2} + 10y(3y)\sqrt[3]{2y^2} = 34y^2\sqrt[3]{2y^2}$

MULTIPLYING RADICALS

There are two types. Sometimes they occur together.

If a, b ≥ 0, then $\sqrt{a} \times \sqrt{b} = \sqrt{ab}$. In fact, if n is positive even, $\sqrt[n]{a} \times \sqrt[n]{b} = \sqrt[n]{ab}$. If n is odd positive, then a and b can be any real number and $\sqrt[n]{a} \times \sqrt[n]{b} = \sqrt[n]{ab}$. In particular, $\sqrt[3]{a} \times \sqrt[3]{b} = \sqrt[3]{ab}$.

Also, $a\sqrt{b} \times c\sqrt{d} = ac\sqrt{bd}$. You multiply the insides and multiply the outsides. Again, both b and d must be greater than or equal to 0.

EXAMPLE 1—

Multiply

$3\sqrt{6} \times 5\sqrt{7}$

The answer is $15\sqrt{42}$.

EXAMPLE 2—

Multiply (the easier way)

$7\sqrt{8} \times 10\sqrt{6}$

$= 70\sqrt{(2)(2)\,(2)(2)\,(3)}$

$= (70)(2)(2)\sqrt{3} = 280\sqrt{3}.$

If you first multiplied 6 times 8, the first thing you would have to do is break it down again. So write each one in terms of primes first to avoid wasting effort.

EXAMPLE 3—

$4\sqrt{2}(3\sqrt{5} + \sqrt{7} - 5 + 8\sqrt{6} + 3\sqrt{2})$

$= 12\sqrt{10} \quad + \quad 4\sqrt{14} \quad - \quad 20\sqrt{2} \quad + \quad 32\sqrt{12} \quad + \quad 12\sqrt{4}$

Multiply outsides Multiply insides	Multiply insides only	Multiply outsides only	Multiply outsides Multiply insides	Multiply outsides Multiply insides
			$\sqrt{12} = 2\sqrt{3}$	$\sqrt{4} = 2$

$= 12\sqrt{10} + 4\sqrt{14} - 20\sqrt{2} + 64\sqrt{3} + 24$

If you couldn't combine inside the parentheses before multiplication, you won't be able to combine terms after the multiplication.

EXAMPLE 4—

Multiply

$(2\sqrt{3} + 4\sqrt{2})(6\sqrt{3} + 2\sqrt{2})$

This looks familiar. Aha!! We must *foil*:

$(2\sqrt{3} + 4\sqrt{2})(6\sqrt{3} + 2\sqrt{2})$

$= 12\sqrt{9} + 4\sqrt{6} + 24\sqrt{6} + 8\sqrt{4}$

$= 36 + 28\sqrt{6} + 16 = 52 + 28\sqrt{6}$

Or, to be fancy,

$4(13 + 7\sqrt{6})$

DIVIDING RADICALS

The last operation is division, but in order to fully do this, we have to rationalize the denominator. Soooo, division will be in the middle of two sections on rationalizing.

Rationalize the Denominator I

Suppose we have $1/\sqrt{2}$. In the past the reason we never wanted radicals in the bottom was the problem of finding an approximate decimal value for $1/\sqrt{2}$. $\sqrt{2}$ is approximately 1.414 (which you should know, as you should $\sqrt{3}$, which is the year George Washington was born—1.732). To find the value of $1/\sqrt{2} = 1/1.414$, we had to divide $1.414\overline{)1.000000}$. This had two great disadvantages: this is very long division, and it is inaccurate because the divisor, 1.414, is rounded off. Now suppose we rationalize the denominator. Look what happens. . . .

EXAMPLE 1

Rationalize

$$\frac{1}{\sqrt{2}} = \frac{1}{\sqrt{2}} \times \frac{\sqrt{2}}{\sqrt{2}} = \frac{\sqrt{2}}{2}$$

By multiplying the top and bottom by $\sqrt{2}$, the arithmetic is much easier to do because . . . if you take $\sqrt{2}/2 = 1.414/2 = 2\overline{)1.414}$, first you are dividing by 2 (very easy) and second, when you round off the top, it is more accurate. However, we really don't have to do this with calculators. In math we almost always have to rationalize the numerator. *Math is behind the times.* I have found that when math is behind the times, math teaching is good. Unfortunately, as I write this book, math is the most progressive subject. In my view, math teaching today is at its worst and so is math learning. This is why I wrote this book!!!! Enough talk. I talk too much. Let's go on!!

EXAMPLE 2—

Rationalize the denominator

$$\frac{7}{\sqrt{75}}$$

Simplify the bottom first!!!

$$\sqrt{75} = \sqrt{3(5)(5)} = 5\sqrt{3}$$

Soooo,

$$\frac{7}{\sqrt{75}} = \frac{7}{5\sqrt{3}} = \frac{7}{5\sqrt{3}} \times \frac{\sqrt{3}}{\sqrt{3}} = \frac{7\sqrt{3}}{15}$$

EXAMPLE 3—

Rationalize the denominator

$$\frac{9}{\sqrt{8a^5}} = \frac{9}{2a^2\sqrt{2a}} = \frac{9}{2a^2\sqrt{2a}} \frac{\sqrt{2a}}{\sqrt{2a}} = \frac{9\sqrt{2a}}{4a^3}$$

Not too bad. Let's do division!!!

Division

Division is based on one simple principle:

$$\sqrt{\frac{a}{b}} = \frac{\sqrt{a}}{\sqrt{b}}$$

where a is not negative, buuuut b must be positive!

Let's do a bunch of problems!!!

EXAMPLE 1—

$$\sqrt{\frac{4}{9}} = \frac{\sqrt{4}}{\sqrt{9}} = \frac{2}{3}$$

Really easy.

EXAMPLE 2—

$$\sqrt{\frac{3}{25}} = \frac{\sqrt{3}}{\sqrt{25}} = \frac{\sqrt{3}}{5}$$

EXAMPLE 3—

$$\sqrt{\frac{7}{27}} = \frac{\sqrt{7}}{\sqrt{27}} = \frac{\sqrt{7} \times \sqrt{3}}{3\sqrt{3} \times \sqrt{3}} = \frac{\sqrt{21}}{9}$$

Notice that once you divide, the problems are just like those in the first part of this section.

EXAMPLE 4—

$$\sqrt{\frac{2a^3b^7}{6a^{10}b^2}} = \sqrt{\frac{1b^5}{3a^7}} = \frac{\sqrt{b^5}}{\sqrt{3a^7}} = \frac{b^2\sqrt{b}}{a^3\sqrt{3a}} \times \frac{\sqrt{3a}}{\sqrt{3a}} = \frac{b^2\sqrt{3ab}}{3a^4}$$

Pretty long, but not bad either. Let's try some more.

EXAMPLE 5—

$$\frac{\sqrt{10a^4}}{\sqrt{35a^7}} = \sqrt{\frac{10a^4}{35a^7}} = \sqrt{\frac{2}{7a^3}} = \frac{\sqrt{2}}{\sqrt{7a^3}} = \frac{\sqrt{2} \times \sqrt{7a}}{a\sqrt{7a} \times \sqrt{7a}}$$

$$= \frac{\sqrt{14a}}{7a^2}$$

We make one square root, simplify, and then break it up.

Last, let's do one with addition.

EXAMPLE 6—

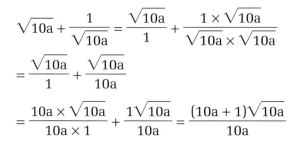

$$\sqrt{10a} + \frac{1}{\sqrt{10a}} = \frac{\sqrt{10a}}{1} + \frac{1 \times \sqrt{10a}}{\sqrt{10a} \times \sqrt{10a}}$$

$$= \frac{\sqrt{10a}}{1} + \frac{\sqrt{10a}}{10a}$$

$$= \frac{10a \times \sqrt{10a}}{10a \times 1} + \frac{1\sqrt{10a}}{10a} = \frac{(10a + 1)\sqrt{10a}}{10a}$$

Rationalize the Denominator II

In the previous parts of this section, when we rationalized the denominator, the bottom was a *monomial* (one term). Let's see what happens when the bottom is a *binomial.*

EXAMPLE I—

Rationalize the denominator

$$\frac{2\sqrt{3}}{4\sqrt{2} + \sqrt{7}}$$

To rationalize, we multiply top and bottom by the *conjugate* of the bottom, $4\sqrt{2} - \sqrt{7}$. To get the conjugate, one changes the sign in the middle.

$$\frac{2\sqrt{3} \times (4\sqrt{2} - \sqrt{7})}{(4\sqrt{2} + \sqrt{7}) \times (4\sqrt{2} - \sqrt{7})} = \frac{8\sqrt{6} - 2\sqrt{21}}{25}$$

Multiply out the top, *foil* the bottom and simplify if possible. The bottom will always be an integer.

EXAMPLE 2—

Rationalize the denominator

$$\frac{\sqrt{c} \times (\sqrt{c} + \sqrt{d})}{(\sqrt{c} - \sqrt{d}) \times (\sqrt{c} + \sqrt{d})} = \frac{c + \sqrt{cd}}{c - d}$$

Note that both terms don't have to be square roots. The conjugate of $6 + \sqrt{x}$ is $6 - \sqrt{x}$. The conjugate of $\sqrt{y} + 3$ is $\sqrt{y} - 3$.

That's all for now. If you've finished reading this whole book and need some algebra review, then read *Algebra for the Clueless.* Algebra, OK? Then you might want *Precalc with Trig for the Clueless.*

Need to study math preparation for the SAT? Read *SAT® Math for the Clueless.*

Good luck. I hope you are really starting to enjoy math.

INDEX

ACKNOWLEDGMENTS

I have many people to thank.

I thank my wife Marlene, who makes life worth living, who is the wind under my wings.

I thank the rest of my family: children and in-law children, Sheryl and Eric, Glenn and Wanda; grandchildren Kira, Evan, Sean, Sarah, and Ethan, one month old at this writing; brother Jerry; parents and in-law parents, Cele and Lee, Edith and Siebeth.

I thank those at McGraw-Hill: Barbara Gilson, John Carleo, John Aliano, David Beckwith, Mary Loebig Giles, Pat Koch, Andrew Littell, Megan McGovern, Maureen Walker, Janice Race, and Nina Levy Girard.

I thank Martin Levine of Market Source for introducing my books to McGraw-Hill.

I thank Daryl Davis, Bernice Rothstein, Sy Solomon, Allysa Dundes, and Dr. Robert Urbanski.

As usual the last thanks go to three terrific people: a great friend Gary Pitkofsky, another terrific friend and fellow teacher David Schwinger, and my sharer of dreams, my cousin Keith Robin Ellis.

ABOUT BOB MILLER...
IN HIS OWN WORDS

I received my B.S. and M.S. in mathematics from Polytechnic University after graduating from George W. Hewlett High School, Hewlett, New York. After my first class, which I taught as a substitute for a full professor at Poly, one student told another upon leaving the room that "at least we have someone who can teach the stuff." I was forever hooked. Since then I have taught at City University of New York (CUNY), Rutgers, and Westfield State College. My name has been included in three editions of *Who's Who Among America's Teachers*.

No matter how badly I feel, I always feel better after I teach. I am always delighted when students tell me they hated math before but now they like it and can do it.

I have a fantastic wife Marlene, tremendous children Sheryl, Eric, Glenn, and Wanda, and unbelievable grandchildren, Kira, Evan, Sean, Sarah, and Ethan. My hobbies are golf, bowling, bridge, and crossword puzzles. My projects for the future are the ultimate high school text, the ultimate calculus text, and a book to teach parents how to teach their young children math, completed and ready for a publisher (anyone interested?). All of these are necessary to ensure that America remains Number 1 forever.

To me, teaching math is a great joy. I hope I can give some of that to you.